CERAMICS AND INORGANIC CRYSTALS
FOR OPTICS, ELECTRO-OPTICS, AND NONLINEAR CONVERSION

Volume 968

CONTENTS

(continued)

PROCEEDINGS

SPIE—The International Society for Optical Engineering

C   nic

C

E

N   n

Rob
Chai

15–1
San

Sponso
SPIE—

Coope
Applied Optics Laboratory/New Mexico State University
Center for Applied Optics Studies/Rose-Hulman Institute of Technology
Center for Applied Optics/University of Alabama in Huntsville
Center for Electro-Optics/University of Dayton
Center for Optical Data Processing at Carnegie Mellon University
Georgia Institute of Technology
Institute of Optics/University of Rochester
Optical Sciences Center/University of Arizona

Published by
SPIE—The International Society for Optical Engineering
P.O. Box 10, Bellingham, Washington 98227-0010 USA
Telephone 206/676-3290 (Pacific Time) • Telex 46-7053

P
Volume 968

SPIE (The Society of Photo-Optical Instrumentation Engineers) is a nonprofit society dedicated to advancing engineering
and scientific applications of optical, electro-optical, and optoelectronic instrumentation, systems, and technology.

The papers appearing in this book comprise the proceedings of the meeting mentioned on the cover and title page. They reflect the authors' opinions and are published as presented and without change, in the interests of timely dissemination. Their inclusion in this publication does not necessarily constitute endorsement by the editors or by SPIE.

Please use the following format to cite material from this book:
    Author(s), "Title of Paper," *Ceramics and Inorganic Crystals for Optics, Electro-Optics, and Nonlinear Conversion,* Robert W. Schwartz, Editor, Proc. SPIE 968, page numbers (1988).

Library of Congress Catalog Card No. 88-62314
ISBN 0-8194-0003-3

CONFERENCE COMMITTEE

Chair
Robert W. Schwartz, Naval Weapons Center

Cochairs
Peter Bordui, Crystal Technology
Saluru B. Krupanidhi, Pennsylvania State University
Donald W. Roy, Coors Ceramics Company

Session Chairs
Session 1—Infrared Transmitting Materials
Donald W. Roy, Coors Ceramics Company

Session 2—Electro-Optics and Nonlinear Conversion I
Saluru B. Krupanidhi, Pennsylvania State University

Session 2 (continued)—Electro-Optics and Nonlinear Conversion II
Peter Bordui, Crystal Technology

Conference 968, *Ceramics and Inorganic Crystals for Optics, Electro-Optics, and Nonlinear Conversion,* was part of a four-conference program on Optical Materials held at SPIE's 32nd Annual International Technical Symposium on Optical & Optoelectronic Applied Science & Engineering, 14–19 August 1988, San Diego, California. The other conferences were

Conference 969, *Diamond Optics*
Conference 970, *Properties and Characteristics of Optical Glass*
Conference 971, *Nonlinear Optical Properties of Organic Materials.*

Program Chair: **Solomon Musikant,** General Electric Company

INTRODUCTION

This proceedings is another in a continuing series dealing with various applications of crystalline optical materials. Session 1 deals with IR transparent materials and includes both the midwave (3-5 μm) and long wave (8-12 μm). The status of these polycrystalline materials range from ready for production, such as spinel, through development and research. Session 2 covers electro-optics and nonlinear conversion. For most of these applications single crystals or glasses are necessary. Materials of particular interest are lithium and potassium niobate, beta-barium borate, and potassium titanyl phosphate.

Robert W. Schwartz
Naval Weapons Center

CERAMICS AND INORGANIC CRYSTALS
FOR OPTICS, ELECTRO-OPTICS, AND NONLINEAR CONVERSION

Volume 968

SESSION 1

Infrared Transmitting Materials

Chair
Donald W. Roy
Coors Ceramics Company

Relationship between atomic structure and selected physical properties of transparent $MgAl_2O_4$

K. E. Green, D. W. Roy

Coors Optical Systems Company
Golden, Co 80401

ABSTRACT

To better understand the basic solid state mechanism affecting select physical properties, structural measurements were made using a relatively new Nuclear Magnetic Resonance
(NMR) technique. Historically, NMR has proved invaluable to the organic researcher for
determining molecular structure(s). For about a decade, exciting work has been ongoing
which has developed and continually refined the technique for solid state NMR. A family
of ceramics known as spinels has a natural crystalline structure which is well defined.
Synthetic spinels do not always adhere to the same strict coordination site scheme. Initial
NMR measurements seem to indicate a possible correlation between the measured coordination
site location(s) of cations and select physical properties.

1. INTRODUCTION

The continuing need for hard, strong, broad transmission bandwidth materials has been
met in recent years by, not a new glass or plastic, but a polycrystalline ceramic. This
material is a specially processed Magnesia-Aluminate spinel ($MgAl_2O_4$). Spinels do occur
naturally or, as in the case of this optical ceramic, are synthesized. Spinels as a
family of materials fall into two structural groups, normal and inverse. The $MgAl_2O_4$ is
a normal spinel in theory and the Al+++ ion is found in octahedral coordination while the
Mg++ is all in Tetrahedral Coordination. An inverse spinel would have a theoretical
structure where all the Mg++ are coordinated octahedrally while half the Al+++ fill the
remaining octahedral sites and the other half go into tetrahedral locations.

It has been mentioned in the literature, that while most naturally occurring spinels
adhere to the theoretic structural scheme, synthetic versions may not. Through recent
advances in NMR techniques, this paper strives to examine relationship between the
measured structure of an optical quality $MgAl_2O_4$ and select physical properties of the
ceramic. The structural determination was done at the regional NMR center at Colorado
State University, Fort Collins, Colorado.

2. SPINEL STRUCTURE

2.1 Theoretical Structure(s)

The polymorphic classification known as spinels encompasses a reasonably large variety
of material. All spinels have the following formula:

$$A++B_2+++O_4$$

The so called "normal" spinels are face centered cubic unit cells with all trivalent
cations (B+++) in octahedral sites and divalent cations (A++) in tetrahedral sites. It
must be pointed out, that there also exists an "inverse" spinel; in this structure, which
is also cubic, the divalent cations occupy the octahedral sites while the trivalent cations
fill the remaining octahedral sites with the balance going into tetrahedral coordination
with the oxygens[1].

2.1.1 $MgAl_2O_4$. Naturally occuring Magnesia-Aluminate spinel ($MgAl_2O_4$) occurs as a normal
spinel structure. This is to say that all Mg++ ions should be in tetrahedral coordination
with the oxygens and all Al+++ should be in octahedral sites.

2.1.2 Synthesis History. A reasonable amount of work has been reported relating to the
laboratory (or industrial) synthesis of $MgAl_2O_4$ powders [2-4]. A further step has been
taken in the literature; and work is reported[5-7], that uses NMR measurements to determine
that synthesized, supposedly "normal" spinels do indeed have non-stoichiometric coordination site ratios. General results reported, indicate that synthesized (as opposed to
natural) $MgAl_2O_4$ powders do not strictly adhere to the proper theoretical coordination
site versus cation charge stoichiometrics as defined for normal spinels.

2.2 NMR Measurements from the Regional Lab. To date, NRM measurements* have been made

*NMR spectras were obtained on a Bruker AM-600 NMR spectrometer CH-1 resonance frequency
of 600 Mhz. All work was done by the Colorado State Regional NMR Center, funded by
National Science Foundation Grant No. CHE-8816437.

on two commercially available $MgAl_2O_4$ powders which are synthesized for further processing into optical spinel. To protect industrial proprieties, sources of the actual powders cannot be disclosed.

A brief and highly simplified description of the NMR spectroscopic technique indicates that any atom with a residual magnetic spin (magnetic moment) when placed in a strong magnetic field (H), will align itself with that field and precess about its own axis with a very specific frequency dictated by quantum mechanical arguments. When the precessing atom is then placed in an RF energy field of the same frequency as the atomic precession (resonant frequency), the RF energy is largely absorbed[8]. This technique is different from other forms of spectroscopy in that it can actually be used to determine atomic (crystalline) structure(s) because the atomic precession frequency is dependent upon (and variable do to) the type, number and bonding of its atomic nearest neighbors!

2.2.1 Measurement Results to Date.

Due to the relative abundance of Al^{27} isotope the measurements pertaining to Al+++ ion coordination have been initially made. Preliminary results are contained in Table 1[9]. Along with other physical property data of interest.

TABLE 1 $MgAl_2O_4$ Powder Properties

Powder	O:T	A:M	BAS(KSI)	MOI(MSI)	KHN	UV
A	7:1	1.04	22.1	39.0	1190	73.4%
B	5:1	1.06	23.6	39.8	1350	70.3%

KEY:
O:T-Octahedral to tetrahedral coordination site ratio for Al+++ ions
A:M-Al_2O_3 to MgO ratio in $MgAl_2O_4$ phase
BAS-Biaxial strength
MOE-Modulus of elasticity (Youngs Modulus)
KHN-Knoop hardness number
UV -Percent ultraviolet energy transmitted through approximately 0.100" at
   0.254 wavelength.

While these data are barely preliminary to this study and have not yet been reproduced, the original indicators point to the NMR measured Al+++ cation site coordinations as being an influence on resultant physical properties. Measured optical transmissivities showed increases as the structure moved toward the theoretical (natural) coordination site ratio.

3. OTHER STRUCTURES

While optical quality spinel ($MgAl_2O_4$) is of increasingly important interest in designing systems where broad transmission bandwidth, non-birefringence, high mechanical strength and hardness (i.e. good wear resistance) are requisite; there are several additional optical ceramic materials.

3.1 Comparisons of $MgAl_2O_4$ and Alternative Optical Materials

The number of polycrystalline materials offered in the past few decades for UV-VIS-NIR optical applications are numerous. Fortunately, the number that find reasonably wide acceptance is smaller and are compared in Table 2.[9-13]

TABLE 2 Comparison of Optical Polycrystalline Ceramics

Material	Crystal Structure	Useable Bandwidth*	Strength	MOE	KHN
$MgAl_2O_4$ (Spinel)	Cubic	0.25-5.2	23 ksi	39 msi	1300
Al_2O_3 (Sapphire)	Hexagonal	0.25-5.1	60 ksi	58 msi	1750
$5AlN \cdot 9Al_2O_3$ (Alon)	Cubic	0.29-4.8	45 ksi	40 msi	1800
Y_2O_3 (Yttria)	Cubic	1.00-6.8	18 ksi	24 msi	800
ZrG_2 (Zirconia)	Cubic	0.41-5.7	16 ksi	28 msi	--
MgF (Irtrani)	Tetragonal	1.00-8.8	12 ksi	17 msi	570

It will be noted here, that while Table 1 is certainly not an all inclusive list of UV thru NIR optical ceramics it does represent the major materials being used in or considered for system(s) designs currently.

4. CONCLUSIONS

At this point in the study relating to effect of polycrystalline optical ceramic(s) properties on crystalline structure as measured by solid state NMR, the only real conclusion one can arrive at is that there appears to be a trend but far more measurements are needed. Salient points to date are:

*Approximate wavelength for 70% transmittance uncoated on samples nominally 1/8" thick.

°Only synthetic $MgAl_2O_4$ has been looked at.
°Only Al+++ cation coordination sites have been determined due to NMR technique constraints relating to Mg++.
°Measurements show that there appear to be the following physical property relationships as the structure approaches theoretical coordination site stoichiometrics.
- optical transmissivities increase slightly
- mechanical properties degrade slightly

5. ACKNOWLEDGEMENTS

The authors are very grateful to and wish to acknowledge the generous assistance of the Colorado State Regional NMR Center at Fort Collins, Colorado working under the auspices of the National Science Foundation Funding Grant No. CHE-8616437. Also to Dr. Steve Dec at the Regional NMR Center who measured and interpreted the spectra and also to Dr. Joel Gray of the Adolph Coors Company, NMR Laboratory, Golden, Colorado for helpful technical liaisons between us and the Regional Center.

8. REFERENCES

1. W. D. Kingery et al, Introduction to Ceramics, 2nd edition, pp. 64-65, John Wiley & Son, New York (1976).
2. H. Okamura and H. K. Bowen, Preparation of Alkoxides for the Sysnthesis of Ceramics, Ceramics International, 12, (1986), pp. 161-171.
3. G. Ritter, Production of Spinel Powder, U.S. Patent No. 4,474,745, Oct. (1984).
4. W. T. Bakker and J. G. Lindsay, Method of Preparing Magnesia Spinel, U.S. Patent No. 3,304,153, Feb.(1967).
5. E. Brun, et al, Electric Quadrafpole change effects of Al^{27} and cation distribution in Spinel ($MgAl_2O_4$), Natturwissenshcaften, 47(12), 277 (1960).
6. E. Brun and S. Hafner, Quadrapole Splitting of Al^{27} in Spinel ($MgAl_2O_4$) and in Corundum (Al_2O_3):I, Nuclear Resonance of Al^{27} and Cation Distribution in Spinel,
7. W. G. Jacobs, Sintering of Reactive $MgAl_2O_4$ Spinel in Controlled Atomspheres, pp. 5-7, PhD Thesis, Rutgers University (1969).
8. G. L. Clark, Encyclopedia of Spectroscopy, pp. 664-674, Reinhold, New York (1960).
9. S. F. Dec, Regional NMR Center, Fort Collins, Colorado, Private Communication, (1988).
10. J. A. Cox et al, Comparative Study of Advanced IR Transmissive Materials, Proceedings SPIE 683 (1986).
11. M. E. Thomas, Infrared Transmission Properties of Sapphire, Spinel, Yttria, and Alon as a Function of Temperature and Frequency, Appied Optics 27(2), pp. 239-244 (1988).
12. J. A. Cox et al, Comparative Study of Advanced IR Transmissive Materials, Proc. SPIE 683 (1986).
13. P. C. Archibald, Optical Measurements on Advanced Performance Domes, SPIE 505, pp. 52-56 (1984).

Lanthana-strengthened yttria domes and windows

G.C.Wei, M.R. Pascucci, E.A. Trickett, C. Brecher, and W.H. Rhodes

GTE Laboratories Incorporated
40 Sylvan Road, Waltham, Massachusetts 02254

ABSTRACT

Lanthana-strengthened yttria is of interest for infrared applications because of its high strength, long wavelength cutoff (8 µm), and low emissivity. Specimens were fabricated for various optical, thermal, and mechanical tests, and their properties measured. A new thermal shock test method was developed and the thermal shock behavior characterized. The toughened yttria was shown to be more resistant to thermal shock fracture than the standard material. Techniques were developed for fabrication of full size infrared domes, and specimens successfully fabricated and characterized. Various details of the material and component properties will be discussed.

1. INTRODUCTION

Lanthana-strengthened yttria (LSY) is one of the most significant new optical materials in the past twenty years [1-6]. Because of its low emissivity, which limits the amount of background radiation generated by heating, this transparent polycrystalline material has a particular advantage for 3-5 µm infrared-transmitting window and dome applications. The low emissivity is a consequence of its long infrared cutoff wavelength ($\approx$ 8 µm), which results in a minimal overlap between the multiphonon absorption band and the blackbody radiosity curve.

In addition to reducing extraneous radiation, the long wavelength infrared cutoff also minimizes the thermal degradation of the desired signal. Since the multiphonon edge of a material gradually moves to shorter wavelengths upon heating, its location at too short a wavelength can allow this absorption to infringe upon the intended region of transmittance. Thus the transmittance of LSY at 3-5 µm is far less sensitive to heating than is the case for most other materials.

To add to the above optical advantages, lanthana-strengthened yttria has other attractive physical attributes. It has cubic symmetry (minimizing scatter loss in polycrystalline samples), a high melting point (2464°C), moderate thermal expansion coefficient (8 x 10^{-6} per °C between 20°C and 1000°C), and satisfactory thermal shock and erosion resistance. Because of all these features, we have sought to exploit this material for IR window and dome applications, with the aim of developing a consistent manufacturing process while reducing the emittance and raising the strength and reliability.

Fabrication of LSY windows and domes starts with high-purity submicron powders. The powders can be pressed into domes, windows, or other complex configurations to near net shape and consolidated to full density through a unique second-phase pressureless sintering technique [7]. This fabrication route provides an economical and rapid process for high-quality optical components, with efficient material usage and economical grinding and polishing costs.

It was discovered that significant toughening and strengthening can be achieved via controlled nucleation and growth of a second phase in the yttria-lanthana system [3]. This is important for improving the resistance of LSY to thermal shock, and rain and dust erosion. Thus, the lanthana additive has been found to play two roles: first, as a sintering aid during consolidation of powder compacts into fully dense, pore-free, transparent bodies; and second, as an agent to control the microstructural development of LSY alloys and thereby impart a strengthening effect.

During the last two years, with the many rapidly developing infrared applications, the need for high-quality large-size LSY windows and domes has continued to grow. In response to this need, full-scale LSY domes were fabricated to demonstrate the material technology. Many test specimens for property measurements were made, and material properties such as scattering, emissivity, absorption, transmittance, thermal expansion, thermal conductivity, specific heat, biaxial flexure strength, modulus of rupture, fracture toughness, tensile strength, Weibull modulus, compressive strength, Young's modulus, and Poisson's ratio were measured. Fundamental characteristics important to the optical and thermal shock performance of LSY in high-temperature high-stress IR-transmitting applications have been ascertained and a large data base for the properties established [8-12]. In order to further characterize the thermal shock behavior, a new thermal shock testing method was developed and utilized in determining the relative resistance to thermal shock fracture for a variety of LSY materials consisting of different microstructures. The high-temperature optical properties are the subject of another paper in this Proceedings [13]. This paper reports the microstructure, room-temperature optical, and room- and high-temperature thermomechanical behavior of lanthana-strengthened yttria.

2. OPTICAL PERFORMANCE

A demonstration of current processing technology is illustrated in Figure 1. These clear transparent domes, 2.8 inch O.D. by 0.075 inch thick, are made of the standard 9 m/o La_2O_3-strengthened Y_2O_3 material with an average grain size of 55 µm. In characterization by the Naval Weapons Center (NWC), the total integrated scattering (TIS) at 3.39 µm for five domes was measured as $3.1 \pm 1.3\%$, better than would be expected for an initial scale-up effort. The high optical quality has been achieved as a result of a thorough understanding of the principles governing each step of the processing.

For flat disks, fabricated in the same manner, an average TIS at 3.39 µm for eight randomly selected specimens was measured as $1.2 \pm 0.6\%$ for 2 mm-thick samples and 1.9% for 6 mm-thick samples. Subsequent to these measurements, the process for fabrication of LSY was further optimized. On representative flat windows (1.3 mm thick) fabricated from the new process, the TIS was measured as low as $0.81 \pm 0.13\%$ at 647 nm and $0.18 \pm 0.03\%$ at 3.39 µm, showing a remarkable optical quality. Because of this low scattering, the transmittance is high and nearly flat throughout the visible and well into the infrared.

Lanthana-strengthened yttria containing lower levels of lanthana than the standard 9 m/o is also under study. A reduction in the lanthana content can, in principle, decrease the magnitude of the calculated thermal stresses induced in LSY domes or windows during transient heating applications. This stress reduction would occur because of the higher thermal conductivity of the low-lanthana material, producing more even temperature distribution and smaller gradients, assuming equivalent mechanical properties.

3. MICROSTRUCTURE

Standard 9 m/o La_2O_3-strengthened Y_2O_3, sintered and annealed, typically exhibits a single-phase pore-free and equiaxed-grain microstructure. The grain size can be tailored by simply adjusting the temperature and duration of sintering and anneal. For example, Figure 2 shows LSY in three different microstructures: single-phase, fine-grained (average grain size = 55 µm); relatively large-grained, single-phase (average grain size = 88 µm) and a fine-grained form containing two distinct phases. The two-phase microstructure was produced by quenching from the two phase (C + H) field of the yttria-lanthana binary system. The fine-grained (<10 µm) microstructure of the two-phase material consists of a mixture of defect-free low lanthana-content cubic phase and high-lanthana-content twinned monoclinic phase, with the latter located either inside the cubic grains or at the grain boundaries [14]. This type of microstructure shows an increase in fracture toughness (1.5 vs 0.9 MPa·m$^{1/2}$) but at the cost of somewhat lower optical transmittance relative to the single phase microstructure [3]. All three types of microstructure were evaluated in their thermal shock behavior as described in the next section. The majority of the thermomechanical property characterization, however, was performed on the standard, single-phase transparent 9 m/o lanthana-strengthened yttria.

4. THERMOMECHANICAL BEHAVIOR

Several thermophysical and mechanical properties are relevant to the performance of 9 m/o La_2O_3-strengthened Y_2O_3 as an IR window or dome. These properties include density, thermal conductivity, heat capacity, modulus of rupture (MOR), elastic modulus, Poisson's ratio, ring-on-ring biaxial flexure strength, tensile strength, compressive strength and Weibull modulus. A large number of test specimens were made for characterization purposes; several test samples from a 55 µm grain size material are shown in Figure 3. The thermomechanical properties have been measured and are listed in Table 1. Some of the results require further discussion.

At room temperature, the MOR values for specimens of the size 25.4 x 2.5 x 1.3 mm ranged from 192 ± 42 MPa to

217 ± 33 MPa, depending on the microstructure of the single-phase alloy. For two-phase toughened LSY bars of the small size, room temperature MOR values were found to be as high as 345 MPa. Clearly, both the grain size and second phase content can affect the mechanical strength significantly. Large MOR bars (101.6 x 12.7 x 6.4 mm) of the standard, single-phase, 9 m/o lanthana-strengthened yttria with a 55 μm average grain size were tested [12] at Southern Research Institute. (SoRI). The MOR values (98.3 ± 9.9, 95.3 ± 13.9, 120 ± 26, and 55.2 ± 6.8 MPa at 25, 540, 1090, and 1650°C, respectively) were about half that measured for the small bars. This trend was reasonable because the large bars had larger surface area and volume under tension than the small bars during the MOR testing.

Fractured surfaces of the large specimens were examined. Most of the specimens tested at room temperature fractured from sites near the corners on the tensile surface. A number of the specimens tested at 540°C and 1090°C fractured from non-corner sites, and several appeared to have fracture origins located not right on the tensile surface, but within the sub-surface tensile volume of the bars. Because of this, elimination of critical volume flaws or defects distributed in the domes or windows has been identified as a material development objective.

Tensile strength was measured at SoRI on cylindrical dog-bone specimens of the standard single-phase 9 m/o lanthana-strengthened yttria with a sizable tensile section (0.156 inch diameter by 1.250 inch long). Absolute tensile strength could not be obtained because all the specimens fractured in the head section instead of the gage section. However, the results did indicate that tensile strength as high as >71.0 MPa (10.3 ksi) was achieved [12], a quite respectable value considering the large surface and volume under tension during the test.

Ring-on-ring biaxial flexure strength of the standard 9 m/o La_2O_3-strengthened Y_2O_3 was measured at room temperature on disks 17.8 mm in diameter by 2 mm thick. The measurements at Southern Research Institute on 40 disks gave average values of 172 ± 34 MPa with a Weibull modulus of 5.4 [12]. Examination of the fractured disks showed that 16 of the 40 disks fractured at locations outside the load ring diameter. The reasons for this are still under investigation, but may involve three sources: (1) surface damage during positioning of the disk on the support ring; (2) disk surfaces that are not completely parallel, which may result in distorted stress profiles; and (3) the relative size of the samples and test fixture, which may lead to stress concentration at or near the sample edges. The average biaxial flexure strength for the group of disks which fractured inside the load ring was 163 ± 37 MPa with a Weibull modulus of 4.9. These Weibull modulus values (4-6) are low compared to those (20-30) of well-developed ceramic products such as alumina and silicon nitride. Further improvements in the strength and Weibull modulus are believed to be achievable through refined processing control.

Fractographic examination of the biaxial flexure strength specimens showed that surface defects on the order of the grain size acted as the failure-initiating flaws [15]. Thus, elimination of such large surface defects is strongly desirable from the standpoint of improving the biaxial flexure strength. Biaxial flexure strength of disks is a good indication of the mechanical strength of domes because the type of stress in biaxial flexure strength specimens simulates those in domes during applications.

The finding that the surface flaws act as fracture origins has important implications. It suggests that Weibull or Batdorf statistical fracture theory involving surface-distributed flaws rather than volume-distributed flaws should be seriously considered for use in thermal-structural calculations of the survivability of the domes or windows for the LSY materials at the present stage of development.

Theoretical parameters were calculated [16] to aid in the assessment of the relative resistance of LSY and four other materials to the initiation and propagation of thermal shock fracture [1]. Such calculations can provide a simplified theoretical comparison of thermal shock resistance, and are much less complicated than the Weibull or Batdorf statistical evaluation of survivability, which require finite element computer calculations of thermal stresses and temperature profiles at successive times. The simplified parameter treatment indicates that LSY ranks high among candidate materials, but shows some inconsistencies that should be resolved experimentally. In this regard, a new thermal shock testing method was developed in order to better characterize the thermal shock performance of a variety of LSY alloys consisting of different microstructures.

When developing an appropriate thermal shock testing method, consideration must be given to the initial state and boundary conditions of the heating process. The thermal stresses developed in the material are controlled by the temperature distribution and thermomechanical properties of the material. The temperature distribution is governed by the initial condition and boundary condition which, in turn depend on the dominant heat transfer mode. The initial state should be identical to that in the actual application because thermophysical and thermomechanical properties of ceramics vary with temperature. Many traditional thermal shock testing methods [17] including (1) quenching bars into water or oil and determining the strength reduction caused by the quenching (the classical Hasselman test), (2) laser thermal shock testing involving rapidly heating the central portions of disk samples irradiated with high power laser, (3) impinging cold-air-jet on hot disks, (4) exposing coated disks to radiation heat flux from high-power heat lamps or high-temperature furnaces, (5) rapidly heating disks in a furnace equipped with focussing mirrors to allow fast heating, (6) quenching bars into fluidized beds, (7) fast heating using an array of solar mirrors, do not satisfy the requirements of both convective-heat-transfer boundary conditions and ambient initial state involved in the transient heating of infrared transmitting domes. The newly developed hot-gas-jet thermal shock testing method, which consists of impinging a stream of high-temperature gas onto the center of a ceramic disk, alleviates the shortcomings of the traditional tests, and appears to be suitable for determining the relative resistance of ceramics to thermal shock involving rapid heating from ambient with the principal heat transfer mode being convection. More detailed discussion of the testing method has been presented[18]. The following is a brief description of the apparatus and method.

The apparatus consists of five parts: heat source (for example, a serpentine heat gun), sample holder, heat source positioning mechanism, heat deflection shutters, and sample temperature measuring technique (Figure 4). The test method consisted of (1) injecting the hot air gun toward the sample, (2) maintaining the hot air gun at a fixed position on top of the sample for 2 min, (3) if the sample does not fracture in the 2 min exposure period, raising the hot air gun to let the sample cool to ambient, and then restarting steps (1) to (2) again at hot air temperatures incrementally higher until the sample fractured, and (4) if the sample fractured, raising the hot air gun and removing the sample when it cooled to ambient. The test results are expressed in terms of (DT_C) which is defined as the critical hot air temperature (the hot air temperature at which the sample fractures) minus the initial temperature (room temperature). The probability of thermal shock fracture (probability of failure) can be calculated by the standard procedure of (1) ranking the DT_C, and (2) calculating the failure probability as the ratio $R/(1 + N)$, where R is the order in the ranking and N the total number of test samples. The feasibility of this testing method has been established using specimens of transparent lanthana-doped yttria. Thermal shock fracture initiates from the cold face of the disk specimens, similar to that typically observed in domes subjected to rapid convective heating.

Using the hot-gas-jet thermal shock testing method, three different microstructures of 9 m/o lanthana-strengthened yttria were characterized for their thermal shock behavior. These three microstructures, shown in Figure 2, are (1) a fine-grained, single-phase microstructure, (2) a relatively large-grained, single-phase microstructure, and (3) a two-phase-toughened microstructure. The results of the thermal shock testing of the first two types are shown in Figure 5. Microstructures consisting of small grains appeared to have high probability of survival for a given critical temperature difference in the test, although the transmittance in the smaller grain size LSY was slightly lower than that of the large grain size LSY (77 vs 79% at 2.5 μm). This shows a trade-off between optical and thermal-shock properties. Figure 6 shows the effect of toughening on the thermal shock behavior. The two-phase-toughened 9 m/o lanthana-strengthened yttria (fracture toughness = 1.5 MPa·m$^{1/2}$) clearly has better thermal shock resistance than the standard single-phase 9 m/o LSY (fracture toughness = 0.9 MPa·m$^{1/2}$). The trade-off between microstructure, optical properties, and thermal shock behavior (using the hot-gas-jet thermal shock apparatus) requires further study. This may prove useful in tailoring the development of a family of LSY materials for specific applications.

<u>5. SUMMARY AND CONCLUSIONS</u>

Our research has demonstrated that lanthana-strengthened yttria is a prime candidate for infrared-transmitting window and dome applications. It has high transmittance to beyond 8 μm, and low emissivity and scattering. Its transmittance remains high at elevated temperatures. It can be fabricated through an economical powder processing technique requiring no pressure for densification, and it has the capacity for toughening and microstructural control.

Significant progress has been made in processing technology and full-size optical quality domes have been fabricated. The optical and thermomechanical properties have been characterized and a large data base has been established. A new testing method for thermal shock survivability has been developed and utilized for the characterization of the thermal shock behavior of LSY. The measurements indicate that the thermal shock resistance can be improved by appropriate tailoring of the microstructure. Overall, the results have brought us to the point where application feasibility has been demonstrated and full implementation requires only mechanical properties refinement and improved process control.

ACKNOWLEDGEMENT

This work was partially supported by the Naval Weapons Center, China Lake, CA, under Contract No. N-60530-86-C-0022.

REFERENCES

1. G. C. Wei, C. Brecher, M. R. Pascucci, E. A. Trickett, and W. H. Rhodes, "Characterization of Lanthana-Strengthened Yttria Infrared Transmitting Materials," *SPIE Proc.* 929 (in press).

2. G. C. Wei, C. Brecher, and W. H. Rhodes, "Effects of Point Defects on High-Temperature Optical Properties in Transparent Polycrystalline La_2O_3-Doped Y_2O_3", *SPIE Proc.* 683, 146-152 (1986).

3. W. H. Rhodes, G. C. Wei, and E. A. Trickett, "Lanthana-Doped Yttria: A New Infrared Window Material", *SPIE Proc.* 683, 12-18 (1986).

4. D. C. Harris and W. R. Compton, "Development of Yttria and Lanthana-Doped Yttria for Infrared Transmitting Domes", presented at the 2nd DOD Electromagnetic Window Symposium, Arnold Air Force Station, TN (October 1987).

5. W. H. Rhodes, "Transparent Polycrystalline Yttria for IR Applications", *Proc. 16th Symposium on Electromagnetic Windows*, Atlanta, GA (June 1982).

6. W.H. Rhodes and E.A. Trickett, "Progress on Transparent Yttria", *Proc. 17th Symposium on Electromagnetic Windows*, Atlanta, GA (July 1984).

7. W. H. Rhodes, "Controlled Transient Solid Second-Phase Sintering of Yttria:, *J. Am. Ceram. Soc.* 64 [1], 12 (1981).

8. M. E. Thomas and W. J. Tropf, "Emission of Polycrystalline Yttria between 500 and 1000°C", The Johns Hopkins University Applied Physics Laboratory Report RL-87-210 (August 1987).

9. R. E. Taylor, H. Groot, and J. Larimore, "Thermophysical Properties of La_2O_3-Doped Y_2O_3, A Report to GTE Labs", Purdue University Report TPRL560 (August 1986).

10. R. E. Taylor, H. Groot, and J. Larimore, "Thermophysical Properties of Y_2O_3-Based Materials, A Report to GTE Labs", Purdue University Report TPRL581 (February 1987).

11. R. E. Taylor, H. Groot, and J. Larimore, "Thermophysical Properties of Two Ceramics", Purdue University Report TPRL581A (November 1987).

12. M. W. Price, T. E. Hubbert, and J. R. Koenig, "Mechanical and Thermal Properties of Four IR Dome Materials," Southern Research Institute Data Report SoRI-EAS-87-1272-6225C (February 1988).

13. C. Brecher, G. C. Wei, W. H. Rhodes, "High Temperature Optical Transmission of Transparent La_2O_3-Strengthened Y_2O_3", this issue.

14. G. C. Wei, T. Emma, W. H. Rhodes, M. P. Harmer, and S. Horvath, "Analytical Transmission Electron Microscopy Study of Phases and Fracture in Y_2O_3-La_2O_3 Alloys," *J. Am. Cer. Soc.* (in press).

15. G. C. Wei, M. R. Pascucci, C. Brecher, E. A. Trickett, and W. H. Rhodes, Properties of Lanthana-Strengthened Yttria Infrared Transmitting Windows," Proc. 19th Electromagnetic Window Symposium, Atlanta, GA, (September, 1988), (in press).

16. D. P. H. Hasselman, "Thermal Stress Resistance Parameters for Brittle Refractory Ceramics, A Compendium," *Bull. Am. Ceram. Soc. 49*, 1033 (1970).

17. D. Lewis, "Thermal Shock Testing of Optical Ceramics," SPIE 297, 120-124 (1981).

18. G. C. Wei and J. Walsh, "Hot-gas-jet Thermal Shock Testing Method and Apparatus", *Am. Cer. Soc. Ann. Mtg. Abstract*, 52 (1988).

19. W. H. Rhodes, J. G. Baldoni, and G. C. Wei, "The Mechanical Properties of La_2O_3-Doped Y_2O_3 ", in GTE Laboratories Technical Report TR 86-818.1 (July 1986).

20. E. A. Trickett, W. H. Rhodes, G. C. Wei, and D. Sordelet, "Infrared Transmitting La_2O_3-Doped Y_2O_3", *Proc. 18th Symposium on Electromagnetic Windows*, Atlanta, GA (September 1986).

Table I
Properties of Single-Phase 9 m/o La_2O_3-Strengthened Y_2O_3

Property		Reference	Property		Reference
Crystal Type:	Fluorite Structure, Cubic Symmetry	[7]	Specific Heat (J/g·K):	0.451 (52°C) 0.501 (202°C) 0.537 (402°C) 0.551 (602°C)	[9]
Density (g/cc):	5.13	[7]			
Melting Point (°C):	2464	[3]	Thermal Expansion Coefficient: 93–889°C, x 10^{-6} C^{-1}	7.7	[12]
Young's Modulus (GPa):	166.5	[19]	Thermal Conductivity (W/m·K):	6.02 (23°C) 4.28 (300°C) 3.99 (500°C) 3.95 (700°C)	[9]
Shear Modulus (GPa):	63.6	[19]			
Bulk Modulus (GPa):	144.5	[19]			
Poisson's Ratio:	0.308	[19]			
Knoop Hardness (kg/mm^2):	730	[5]	Refractive Index:	1.9699 (546.1 nm) 1.8703 (5 µm)	[2]
Four-Point Flexural Strength (MPa): Specimen Size: 25.4 x 2.5 x 1.3 mm	217 ± 33	[15]	Temperature Coefficient of Refractive Index (x 10^{-6} C^{-1}), 20° to 1000°C	50 (0.357 µm) 32 (3.29 µm)	[2]
Ring-On-Ring Biaxial Flexural Strength (MPa): Specimen Size: 17.8 mm dia. x 2 mm Thick	172 ± 34	[12]			
Weibull Modulus:	5.4	[12]	Absorption Coefficient (cm^{-1}):	0.01 (3 µm) 0.01 (4 µm) 0.01 (5 µm)	[1]
Fracture Toughness (MPa · m$^{1/2}$):	0.9 – 1.5	[19]			
Total Integrated Scattering (%):	1.19 ± 0.60 (3.39 µm)	[4]			
Dielectric Constant:	12.2 (100 kHz)	[20]	Temperature Dependence of Absorption Coefficient (cm^{-1}) at 3.39 µm:	Almost Constant from 20° to About 900°C	[2]
Dielectric Loss Tangent:	<0.0001 (100 kHz)	[20]			
Wavelength Range (µm) for Transmission >50%, at 2.5 mm Thickness	7.9	[2,4,5]	Emissivity at 5 µm:	0.01 (23°C) 0.04 (500°C) 0.10 (750°C) 0.20 (1000°C)	[8]

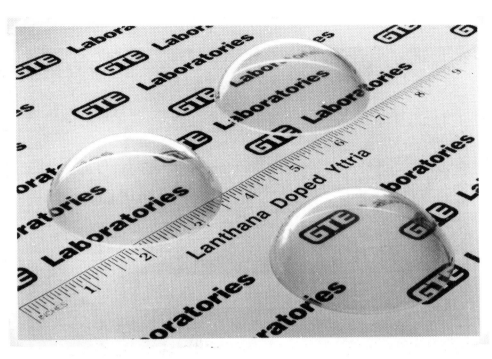

Figure 1. Full-size lanthana-strengthened yttria domes illustrating near-net-shape fabrication and high level of optical quality.

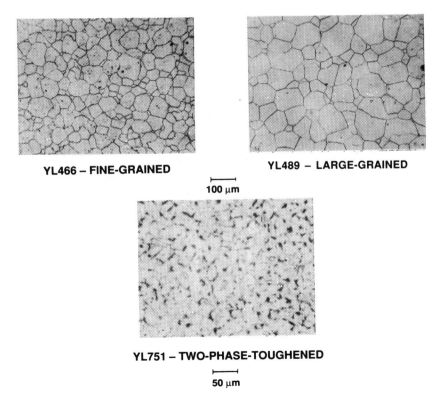

YL466 – FINE-GRAINED

YL489 – LARGE-GRAINED

100 μm

YL751 – TWO-PHASE-TOUGHENED

50 μm

Figure 2. Microstructures of Y_2O_3-La_2O_3 discs subjected to the hot-gas-jet thermal shock test.

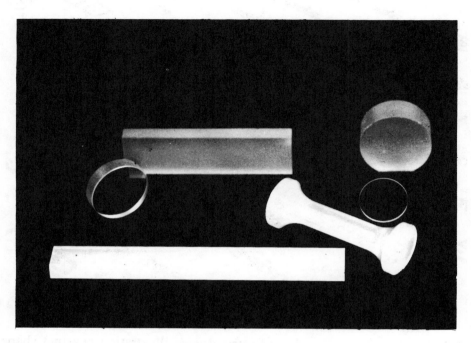

Figure 3.   A representative collection of optical, thermal, and mechanical specimens of lanthana-strengthened yttria fabricated for property testing.

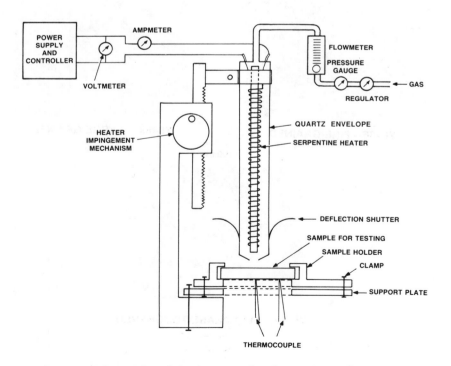

Figure 4.   Schematic of the hot-gas-jet thermal shock apparatus.

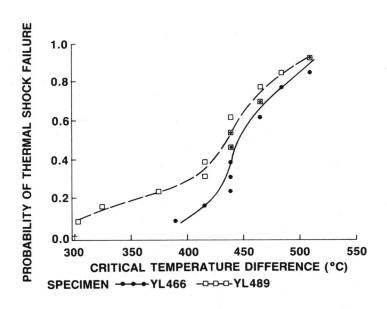

Figure 5. Thermal shock testing behavior of lanthana-strengthened yttria vs grain size.

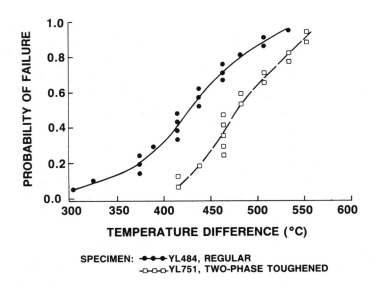

Figure 6. Thermal shock testing behavior of lanthana-strenthened yttria vs toughening.

High-Temperature Optical Transmission of Transparent La_2O_3-Strengthened Y_2O_3

C. Brecher, G. C. Wei, and W. H. Rhodes

GTE Laboratories Incorporated, 40 Sylvan Road, Waltham, MA 02254

ABSTRACT

The optical transmission of transparent polycrystalline lanthana-strengthened yttria has been measured in both the near ultraviolet and infrared regions at temperatures between 20°C and 1400°C. The absorption remains extremely low until about 900°C, then rises almost exponentially as the temperature is raised further. The magnitude of the increase is a function of the oxygen partial pressure (P_{O_2}) in the firing atmosphere. The absorption increase with temperature is smallest when P_{O_2} is between 10^{-11} and 10^{-8} atm, representing the range in which the concentration of stoichiometry-related point defects (oxygen interstitials and vacancies) is minimized. The temperature dependence is significantly greater in the UV than in the IR, but the optimal P_{O_2} range is the same. The absorption behavior is also a function of processing variations, and provides a good criterion for comparison of specimen quality.

INTRODUCTION

Lanthana-strengthened yttria is a material with great potential for optical window applications [1-4]. It has high transmissivity at wavelengths from 300 nm to 8 µm [1]. The long wavelength cutoff and low absorption and scattering result in low emissivity, which allows more tolerance for heating than other IR-transmitting oxides, without interference from phonon edge shift. The material has high strength and toughness, and can be fabricated into domes and windows to near net shape by economical powder processing techniques.

The optical properties of this material have been studied at room and elevated temperatures as functions of processing variations. It was found that a post-sintering anneal treatment plays a major role in maximizing the transmittance at room temperature and removing extraneous absorptions due to OH⁻ contamination. The transmittance, and particularly the temperature dependence of this transmittance, were found to be extremely sensitive to the partial pressure of oxygen in the atmosphere in which the specimens are annealed. The transmission loss and its temperature dependence reach a minimum in the same range of oxygen partial pressures throughout the entire useful wavelength range, suggesting a common origin. The behavior in the ultraviolet has already been discussed in a previous paper [1], but without correlation with the infrared. The behavior in the latter spectral region is the focus of the present work.

EXPERIMENTAL

Specimens of lanthana-strengthened yttria containing 9 ^{m}/o La_2O_3 were prepared from high-purity submicron powders by pressing into disks and consolidating to full density through a unique transient second phase sintering technique [5]. The specimens were annealed at 1400°C in a flowing gas stream containing various partial pressures of oxygen, to adjust stoichiometry and expel OH⁻ contamination. The specimens were ground and polished into the form of truncated semicircular disks about 2 cm in diameter and 2 or 6 mm thick. The specimens were clear and transparent, with high broadband transmissivity from about 300 nm in the near ultraviolet to about 8 µm in the infrared. The material has an electronic transition edge at shorter wavelengths in the UV, while the long-wavelength transmission limit is defined by the multiphonon absorption edge in the IR. The sharpness of the edges and the flatness of the transmitting range can be affected by processing-related factors, such as scattering from optical inhomogeneities, UV and visible absorption by point defects, and IR absorption bands from impurity ions. The most prominent of these is an absorption at about 3 µm caused by OH⁻ contamination [6,7]. This absorption, which has a characteristic two-component structure (see Figures 2 and 5 of Reference 6), can severely impact the transmission at 3.39 µm. Fortunately, this absorption can be readily removed by the post-fabrication annealing treatment [6]. Scattering, although present throughout the optical range, is most evident at the shorter wavelengths. In our specimens, however, the flatness of the transmission (Figure 1) indicates a remarkably low degree of such scattering; this is confirmed by direct measurement of total integrated scattering [3], which gives values as low as 0.18±0.03% at 3.39 µm and 0.81±0.13% at 647 nm, a clear demonstration of the high optical quality of the material.

The optical measurements are made through a double beam technique, which is largely insensitive to errors that may arise from power fluctuations, surface artifacts, and specular reflection from the surface. Two matched specimens, identical except for their thickness, are inserted into an alumina tube closed with optical windows, which is in turn enclosed within a globar furnace capable of heating the samples to temperatures in excess of 1400°C. The measurement technique involves splitting the output from a suitable laser (in the infrared, 3.39 µm He-Ne) into two beams and passing them through the specimens. The beams are then directed into matched detectors and the ratio of their outputs is recorded. The optical system is depicted schematically in Figure 2.

The absorption coefficient is readily calculated from the ratio of the transmittance values of two specimens differing only in thickness. The relationship can be expressed by the equation

$$T_1/T_2 = \frac{1-r^2(1-s)^2\exp(-2\alpha x_2)}{1-r^2(1-s)^2\exp(-2\alpha x_1)} \cdot \exp(-\alpha(x_1-x_2)) \ , \tag{1}$$

where $T = I/I_0$ is the transmittance of each specimen,
         $r = (n-1)^2/(n+1)^2$ is the reflectivity at each surface,
         s   is the fractional scattering loss at each surface,
         x   is the sample thickness of each specimen in cm,
and α is the absorption coefficient in cm^{-1}.

This last term, to be precise, includes not only absorption but also losses due to distributed scattering throughout the bulk, which follows the same functional form. However, as indicated earlier, the latter is small in the infrared, and is also unlikely to show significant temperature dependence.

The particular advantage of this double beam technique lies in its ability to remove the effect of losses that occur at the surface. The theoretical reflectivity at normal incidence can be calculated directly from the refractive index, whose room temperature values and temperature coefficients are known [1]. The surface scattering term, in direct measurements on similar samples [3], was found to be at least an order of magnitude less than the reflectivity; having only minimal effect ($<0.1\%$) on the ultimate solution, this term can safely be neglected. The calculation itself involves an iterative computerized solution to Equation (1), but convergence is rapid and the roots unique. The results are summarized in Table I.

RESULTS AND DISCUSSION

A series of measurements was performed to define the effect of oxygen partial pressure in the annealing atmosphere upon the high-temperature optical transmission of yttria-lanthana at 3.39 µm. The measurements covered a P_{O_2} range from 10^{-16} to 10^{-3} atm, and temperatures from 20°C to 1400°C. The results of a study in the near ultraviolet (357 nm) have already been reported [1]. The high-temperature (1400°C) UV absorption coefficients were found to fall in the range of 11.5 to 13 cm^{-1}, a factor of about 100 greater than at room temperature (Figure 3). In the infrared (3.39 µm), however, we find that the transmission losses and their temperature dependence are much lower than was the case in the near ultraviolet. The average room temperature absorption coefficient at 3.39 µm is about a factor of five lower, and at 1400°C as much as a factor of 30 lower, than the corresponding value at 357 nm. In most cases, the high-temperature infrared absorption coefficient does not reach 0.5 cm^{-1}, and even in the worst cases (the extremes of the O_2 partial pressure range) barely exceeds twice that value.

The dependence of the absorption coefficient on temperature, in various oxygen partial pressures, is shown in Figure 4. At the intermediate P_{O_2} values there is a common pattern of behavior, with two distinct temperature regions. The absorption coefficient is quite low at room temperature, and increases relatively slowly below about 900°C. Above that point, however, the rate of increase with temperature accelerates substantially, ultimately approximating an exponential shape. At either end of the P_{O_2} range the behavior is similar but more extreme. At the lowest oxygen partial pressure, 10^{-16} atm, the absorption at room temperature is double the average of the others, but it has a higher threshold for rapid increase (≈ 1000°C), and a greater rate of increase thereafter. At the highest P_{O_2} value (10^{-3} atm), in contrast, the absorption coefficient shows by far the greatest rate of increase at the lower temperatures, with no clear threshold for acceleration.

The dependence of the absorption coefficient on oxygen partial pressure is shown in Figure 5. At the lower temperatures, the lowest transmission losses are found at P_{O_2} values considerably higher (10^{-6} to 10^{-8} atm) than would have been expected from the UV measurements. The optimum, however, is quite broad, and appears to shift to lower O_2 pressures as the temperature is raised, so that by 1400°C it occurs at about 10^{-11} atm P_{O_2}, not too far from the approximately 10^{-12} atm P_{O_2} value found in the UV. These results suggest the choice of annealing atmosphere at about 10^{-10} to 10^{-11} atm P_{O_2} to achieve the optimum high-temperature transmittance.

The temperature-dependent variation of the absorption is associated with the presence of point defects due to local departures from stoichiometry, i.e., oxygen interstitials and vacancies [1,8]. The yttria lattice, in which only three-quarters of the potential anion sites are actually occupied, is particularly susceptible to such defects. Oxygen interstitials are readily generated under oxidizing conditions, while oxygen vacancies are produced in reducing environments. These point defects provide local levels within the 5.5 eV band gap of Y_2O_3. These levels cause strong absorptions in the near UV, with lower energy tails extending through the visible into the infrared, where they merge into the high-energy tail of the multiphonon edge. Since the shift of the multiphonon edge with temperature is relatively small, it must be the point-defect-induced absorption process that is primarily responsible for the temperature-dependent rise in absorption at 3.39 µm. The fact that this increase, in absolute terms, is smaller than in the UV is not at all surprising, since the temperature dependence in the infrared tail would be expected to be weaker than that of the main absorption. On the other hand, the dependence on P_{O_2} is even stronger in the IR, where the absorption at 1400°C ranges over a factor of three (compared to 15% in the UV). Overall, the results of the measurements at the two wavelengths are fully consistent with a common origin.

Additional information on the nature of the point-defect-induced absorption mechanism can be obtained by examining the relationship between the absorption coefficient increase and the oxygen partial pressure. According to theoretical considerations [8], the equilibrium concentration of various point defects is proportional to different powers of the oxygen concentration, depending upon the specific nature of the defect. Thus, the slope of a log-log plot of thermally activated absorption coefficient vs P_{O_2} should give some indication of the nature of the dominant point defects. When we do this with our data at 1400°C (Figure 6), we find in the strongly reducing range ($P_{O_2} < 10^{-12}$ atm) a slope of about -1/6.4, close to the theoretical slope of -1/6 for doubly charged oxygen vacancies. Similarly, the slope in the strongly oxidizing range ($P_{O_2} > 10^{-5}$ atm) suggests singly charged oxygen interstitials as the major type of point defect. These considerations are valid under the assumption that intrinsic (thermally activated) defects dominate in the material; if extrinsic (impurity-related) defects dominate, the theoretical slopes in the two ranges should be -1/4 and 1/4, respectively. Although the good correlation gives additional support to the hypothesis of a point defect mechanism, one should resist the temptation to read too much into the actual values, particularly in view of the sparseness of experimental points. Furthermore, it should be noted that these slopes are distinctly temperature dependent, especially in the reducing range, suggesting the participation of more than one kind of defect.

The question remains as to whether the point-defect-induced rise in infrared absorption with temperature is intrinsic or extrinsic. One relevant point is the broad range of P_{O_2} in which both the UV and the IR absorptions are near their minima. This suggests an extrinsic origin for the absorption, with the concentrations of the point defects largely controlled by aliovalent impurities. If the centers were intrinsic to the material, the dependence on oxygen partial pressure should be much stronger, with a narrower minimum whose location is determined by the thermodynamics of the point defects. Even in this case, however, an ordering or clustering of oxygen interstitials or vacancies could significantly broaden the optimal P_{O_2} range. Earlier results of weight gain measurements [9] obtained by oxidation of Y_2O_3-La_2O_3 specimens showed the net concentration of oxygen interstitials to be on the order of 50 ppm, the same as the estimated total concentration of aliovalent impurities. While not conclusive, the preponderance of evidence does indicate extrinsic behavior.

CONCLUSION

This work has characterized the temperature dependent behavior of the absorptive loss of lanthana-strengthened yttria optical windows at 3.39 μm. The absorption coefficient at room temperature is low (≤ 0.1 cm^{-1}). The loss increases with temperature, slowly at first but at an increasingly more rapid rate as the temperature exceeds about 900°C. The absorption and its rate of increase with temperature are functions of the oxygen partial pressure under which the specimen had been annealed, and are minimized at a P_{O_2} of about 10^{-11} atm. The absorption in the infrared region, as in the ultraviolet, appears to originate from point defects in the lattice. Questions regarding the spatial distribution of the defects, such as clustering and/or association with aliovalent impurities, remain open.

ACKNOWLEDGMENT

This work was partially supported by the Naval Weapons Center, China Lake, CA, under Contract No. N-60530-86-C-0022.

REFERENCES

1. G. C. Wei, C. Brecher, and W. H. Rhodes, "Effects of Point Defects on High-Temperature Optical Properties in Transparent Polycrystalline La_2O_3-Doped Y_2O_3", *SPIE Proc.* **683**, 146-152 (1986).

2. W. H. Rhodes, G. C. Wei, and E. A. Trickett, "Lanthana-Doped Yttria: A New Infrared Window Material", *SPIE Proc.* **683**, 12-18 (1986).

3. D. C. Harris and W. R. Compton, "Development of Yttria and Lanthana-Doped Yttria for Infrared Transmitting Domes", presented at the *2nd* DoD Electromagnetic Window Symposium, Arnold Air Force Station, TN (October 1987).

4. W. H. Rhodes, "Transparent Polycrystalline Yttria for IR Applications", *Proc. 16th Symposium on Electromagnetic Windows*, Atlanta, GA (June 1982).

5. W. H. Rhodes, "Controlled Transient Solid Second-Phase Sintering of Yttria", *J. Am. Ceram. Soc.* **64**[1], 13-19 (1981).

6. G. C. Wei, "Extrinsic OH$^-$ Absorption in Transparent Polycrystalline Lanthana-Doped Yttria", *Comm. Am. Ceram. Soc.* **71**[1], C20-23 (1988).

7. J. A. Harrington and C. Greskovich, "Infrared Absorption in ThO_2-Doped Y_2O_3", *J. Applied Phys.* **48**[4], 1585-88 (1977).

8. H. Tsuiki, T. Masumoto, K. Kitazawa, and K. Fueki, "Effect of Point Defects on Laser Oscillation Properties of Nd^{3+}-Doped Y_2O_3", *Jpn. J. Appl. Phys.* **21**[7], 1017-21 (1980).

9. W. H. Rhodes, E. A. Trickett, and G. C. Wei, "Processing Studies for Optically Transparent La_2O_3-Doped Y_2O_3", *GTE Laboratories Technical Report* **TR 85-818.1**, Annual Report on Office of Naval Research Contract N00014-82-C0452 (July, 1986).

Table I. Dependence of Absorption Coefficient of Y_2O_3-La_2O_3 at 3.39 μm upon Temperature and Oxygen Partial Pressure.

Temperature (°C)	Absorption Coefficient (cm⁻¹) at				
	P_{O_2} = 10^{-16}	10^{-12}	10^{-8}	10^{-5}	10^{-3} atm
20	0.1968	0.1271	0.0797	0.0721	0.0629
200	0.2161	0.1377	0.0883	0.0788	0.0917
300	0.2268	0.1478	0.0929	0.0861	0.1178
400	0.2374	0.1587	0.0973	0.0933	0.1519
500	0.2479	0.1692	0.1038	0.1031	0.1914
600	0.2612	0.1824	0.1123	0.1137	0.2359
700	0.2713	0.1974	0.1219	0.1279	0.2849
800	0.2782	0.2143	0.1326	0.1431	0.3385
900	0.2879	0.2292	0.1484	0.1587	0.4009
1000	0.3031	0.2496	0.1700	0.1821	0.4745
1100	0.3265	0.2710	0.2021	0.2190	0.5684
1200	0.3807	0.3013	0.2502	0.2753	0.6941
1300	0.5740	0.3396	0.3130	0.3624	0.8747
1400	1.1212	0.3910	0.4032	0.4869	1.1440

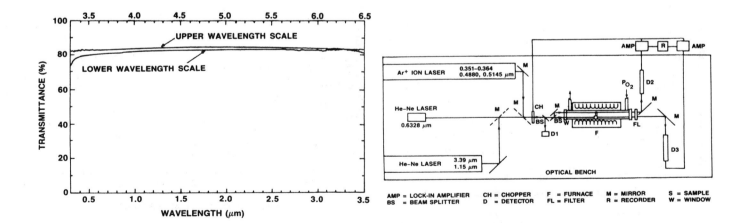

Figure 1. Transmittance of yttria-lanthana in the visible and infrared.

Figure 2. Schematic of optical system for laser transmittance measurements at elevated temperatures.

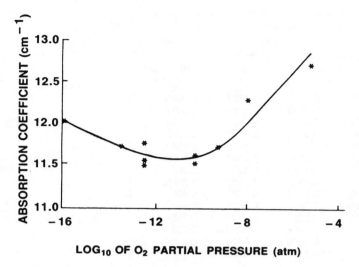

Figure 3. High-temperature (1400°C) absorption coefficient of yttria-lanthana at 357 nm, as function
of oxygen partial pressure.

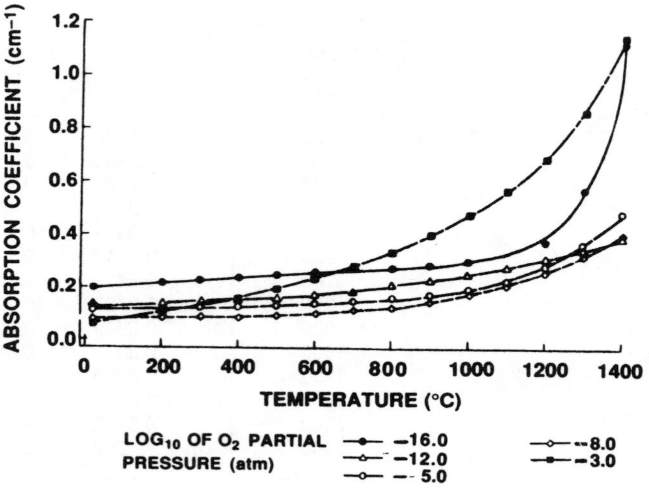

Figure 4. Absorption coefficient of yttria-lanthana at 3.39
μm as function of temperature, after firing at
various O_2 partial pressures.

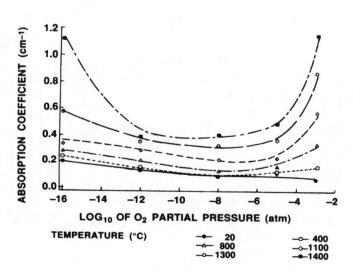

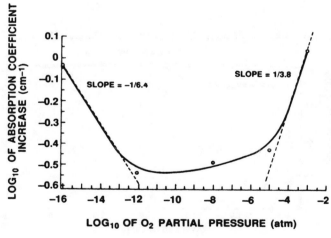

Figure 5. Absorption coefficient of yttria-lanthana at 3.39
μm as function of oxygen partial pressure, at
various specimen temperatures.

Figure 6. Log_{10} of thermally activated absorption coefficient of yttria-lanthana at 3.39 μm, after firing
in various O_2 partial pressures. Values represent increase in magnitude between 20°C and
1400°C.

Optical, mechanical and thermal properties of Calcium Lanthanum Sulphide compared
with the properties of current 8-12 μm window materials

J A Savage and M J Edwards

Royal Signals and Radar Establishment, St Andrews Road, Malvern, Worcs WR14 3PS

ABSTRACT

The physical property data reported for calcium lanthanum sulphide indicates that it would be an attractive material for high performance 8-12 μm FLIR window applications. However, a number of other available materials may also be considered for aircraft windows operating at temperatures up to 200°C. These are low resistivity germanium (70°C to 100°C), gallium arsenide, multi spectral and standard grade zinc sulphide, zinc sulphide/zinc selenide laminate and zinc selenide. A comparison is presented of the transmittance properties of these materials at room temperature and at 175°C together with room temperature data on the reported optical, mechanical and thermal properties. A qualitative assessment of the likely effectiveness of calcium lanthanum sulphide as a FLIR window material in relation to these existing materials is presented.

1. INTRODUCTION

Studies have been carried out to identify 8-12μm materials likely to possess improved physical properties over those of zinc sulphide and rare earth ternary sulphides have been shown to be leading candidates for investigation. Of these, calcium lanthanum sulphide has received the most attention and research on this material has reached a point where a development programme to provide pilot production technology could be contemplated if required. It is therefore timely to consider the possible future role of this material in relation to other available or potentially available materials.

The physical property data reported for calcium lanthanum sulphide indicates that it would be an attractive material for high performance 8-12 μm FLIR window applications. It demonstrates an improved water drop impact performance and an extended IR transmittance range over those of as grown zinc sulphide. However, a number of other available materials may also be considered for aircraft windows operating at temperatures up to 200°C. These are low resistivity germanium (up to 70° to 100°C), gallium arsenide, multispectral grade zinc sulphide, zinc sulphide/zinc selenide laminate and zinc selenide. Since much physical property data on calcium lanthanum sulphide has now been assembled, it is useful to compare this with similar property data for the above materials to see whether there would be any advantages in using this new material for aircraft window applications.

2. SYNTHESIS OF CALCIUM LANTHANUM SULPHIDE

Rare earth ternary sulphides of the general formula AB_2S_4 where A is a divalent cation and B is a trivalent rare earth cation crystallize amongst six structural families (Flahaut 1979). Isotropic materials are preferred for window applications so that cubic spinel and cubic thorium phosphide structures are of interest. Of the rare earth cations lanthanum and yttrium are to be preferred on cost and abundance criteria and this leads to interest in thorium phosphide compounds containing lanthanum and spinel compounds containing yttrium. Data on the infrared absorption frequencies (Provenzano 1976) indicates that thorium phosphide compounds are to be preferred to gain the maximum transmittance range. Of the divalent cations forming cubic structures with lanthanum, Ca is the smallest and thus may be expected to offer the highest strength properties when incorporated in the compound $CaLa_2S_4$. Hence, amongst rare earth ternary sulphides work has concentrated on calcium lanthanum sulphide for IR window applications. In the light of recent physical property data published for a number of rare earth ternary sulphides (Chess et al 1984) this has proved to be a good choice.

Work on calcium lanthanum sulphide has shown (Saunders et al 1986) (Walker and Ward 1984) that there is a region of solid solution in the Ca-La-S phase diagram between $CaLa_2S_4$ and La_2S_3. Deviation from this join to yield material with increased or decreased sulphur content is likely to have detrimental effects on the physical properties. For instance reduced material can become semiconducting or metallic. In addition, since there are a number of reports of CaS inclusions in both raw material powder and ceramic (Beswick et al 1983) (Covino et al 1984), La_2S_3 rich material is being investigated in relation to physical properties (Saunders et al 1986) (Savage and Lewis 1986). The most successful technique for the fabrication of optically transmitting specimens has been reported to be a ceramics route starting from fine grain sulphide powder followed by sintering of green bodies made from this powder and finally Hipping of the sintered material. Mixed oxide powder made through a carbonate precipitation route or an evaporative decomposition of droplets of mixed nitrate solution route is converted to sulphide powder by firing at 1000°C in H_2S gas for 12-24 hours. Green bodies of ≥ 55% of theoretical density are made by hot pressing or cold pressing with a binder and then are sintered in H_2S gas in the temperature range 1050°C to 1200°C for 5 to 6 hours to form ceramic bodies of ≥ 95% of theoretical density. These bodies are then hot isostatically pressed (HIP) in the temperature range 990°C to 1200°C for 15m to 3h at 172 to 200 MPa to yield optically transparent ceramics. Studies have been carried out of the physical properties of calcium lanthanum sulphide in relation to the CaS/La_2S_3 ratio (Saunders et al 1986) (Savage and Lewis 1986) (Harris et al 1987) (Lewis and Beswick 1985) (Lewis et al 1988) and therefore much data is available. This data is not representative of a unique composition but has been mainly measured on compositions in the range 40 CaS: 60 La_2S_3 to 50 CaS: 50 La_2S_3. The physical properties change relatively slowly with composition within this range and thus the data are considered sufficiently acceptable for the present comparison. Most of the data of table 1 has been taken for 42.5 CaS: 57.5 La_2S_3 composition with gaps filled where necessary with data for 50 CaS: 50 La_2S_3 composition.

3. PHYSICAL PROPERTY COMPARISON

There is a need to deploy 8-12 μm thermal systems in the air environment inside fixed wing aircraft. For this application robust windows are required to protect these systems during high speed flight which is likely to generate high window temperatures. Thus materials with extended IR transmittance range are to be preferred to avoid excessive absorption/emittance at the long wavelength end of the window band pass. On examination of table 1 and figure 1, it can be seen that there are no ideal materials in terms of thermal, mechanical and far IR transmittance range properties. Thus systems performance compromises may need to be made to allow for the optical and the thermo-mechanical limitations of available materials such as germanium, gallium arsenide, standard grade ZnS, multi spectral grade ZnS, ZnSe and potentially available materials such as ZnS/ZnSe laminate and calcium lanthanum sulphide. Each application will need to be considered separately in terms of system and environmental parameters as well as window material properties, thickness and coatings performance. It is appropriate for the present purposes to compare the properties of the materials in a relative manner.

3.1 Optical properties

The refractive index data of Table 1 is taken from manufacturers literature or measurements performed at the NPL UK on Littrow prisms. The far IR transmittance ranges of some 8-12 μm window materials are compared in figure 1 for a 1mm thickness at 20°C and at 175°C (100°C for Ge) using a Perkin Elmer 983 spectrophotometer equipped with a Specac variable temperature cell. The different levels of transmittance are in relation to the different levels of refractive index of the materials. Germanium offers useful performance out to 12 μm with some multi phonon absorption occuring between 11 and 12 μm due to an intrinsic absorption band at 847 cm^1. This performance is achieved up to about 70-100°C preferably with low resistivity material but beyond this temperature there are serious losses due to free electron absorption. GaAs offers useful performance out to 12 μm up to 200°C although there is likely to be some loss due to the phonon band at 760-770 cm^{-1} at the far infrared end of the 8-12μm spectral region in realistic window thicknesses. However, it is an expensive material and the lack of visible and very near IR transmittance may be a problem for some applications. ZnS is intended as a high temperature airborne window material, is relatively cheap and transmits through from the visible to the far IR (multispectral material) or mainly in the 8-12 μm band (as grown CVD material). However, the phonon absorption band at 900 cm^{-1} is likely to limit the usefulness to the 8-10 μm band and the absorption/emittance may be a problem at 10 μm for some applications requiring thick windows operating near 200°C. Calcium lanthanum sulphide offers an 8-12 μm performance close to that of GaAs with some absorptance/emittance likely to occur in the 11-12 μm region for realistic window thicknesses. ZnSe offers the best performance of all for 8-12μm applications even for very thick windows of the order of 14mm as reported by Klein et al (1986) in a comparison of ZnS, ZnSe and ZnS/ZnSe laminate for FLIR applications. The ZnS/ZnSe laminate is an improvement optically over ZnS in the 8-10 μm band (18mm thickness) at 200°C. It is not represented in Table 1 or figure 1 since measurements would only be applicable for thicker material ie 1mm of ZnS plus several mm of ZnSe. The inline transmittance values taken at 3 wavelengths and two temperatures are listed in table 1 for all of the 1mm thick samples of figure 1. It is clearly seen that the transmittance values at 12 μm are affected by increase in temperature for most of the materials even at a thickness of 1mm and that calcium lanthanum sulphide offers an improvement over ZnS.

3.2 Mechanical properties

From the data listed in table 1 it can been seen that the hardness of calcium lanthanum sulphide is greater than that of FLIR grade ZnS or ZnSe while the fracture toughness on early samples approaches the values for ZnS and ZnSe. Similarly early rupture modulus measurement values for calcium lanthanum sulphide approach those of ZnS and germanium and are likely to equal them when material with a similar state of perfection is developed. The Young's modulus of calcium lanthanum sulphide is higher than that of ZnS and this is reflected in the improved water drop damage threshold listed in Table 1. On balance calcium lanthanium sulphide is likely to offer similar basic strength properties to Ge and ZnS but with an improved rain erosion resistance.

3.3 Thermal properties

The basic material properties, the window geometry, the flight profile and the mounting are all important parameters influencing the thermal performance of an aircraft window. Consideration of all of these parameters is beyond the scope of this work so that comment will be limited to the influence of material properties on thermal shock. The thermal expansion coefficient of calcium lanthanum sulphide is high and the thermal conductivity is low compared with the other materials listed in table 1. These values are likely to be acceptable where very sudden thermal shock is not a problem. Hasselman (1970) has established a thermal shock figure of merit for the comparison of different materials. The parameter, R', applies to mild thermal shock;

$$R' = \sigma k (1-\nu)/\alpha E$$

where σ = fracture strength, ν = Poisson's ratio, k = thermal conductivity, α = thermal expansion coefficient and E = Young's modulus. It can be seen that germanium and gallium arsenide offer the best thermal shock performance with ZnS being superior to calcium lanthanum sulphide. For comparison purposes the values for two 3-5 μm materials are MgF$_2$ 0.89 and sapphire 2.1 (Gentilman 1986).

4. CHOICE OF MATERIAL

It is clear from the above that there is no ideal material for aircraft window requirements. Germanium offers a useful optical/mechanical/thermal performance for window temperatures up to 70-100°C and has seen considerable service. For operational temperatures nearer to 200°C then there is a choice between GaAs, ZnS or ZnSe available as off the shelf materials. In relation to Ge, GaAs is more expensive and less strong while ZnS offers equivalent strength but worse optical transmittance range with the thermal properties likely to be accpetable for most applications. ZnSe offers the best optical performance but the worst mechanical performance and is again expensive. Calcium lanthanum sulphide would offer a useful optical/mechanical performance and might prove to be a suitable alternative to ZnS if the thermal performance were acceptable.

ACKNOWLEDGEMENT

Thanks are due to Mr W R MacEwan for supply of the GaAs sample.

REFERENCES

Beswick J. A., Pedder D. J., Lewis J. C. and Ainger F. W., Proc. SPIE 400, 12-20 (1983).
Blair P. W., Thesis, Cambridge University UK. The liquid impact behaviour of composites and some IR transparent materials (1981).
Chess D. L., Chess C. A., Biggars V. J. and White W. B., J. Amer. Ceram. Soc. 66, 18-22 (1983).
Chess D. L., Chess C. A. and White W. B., Mat. Res. Bull. 19, 1551-1558 (1984).
Covino J., Proc. SPIE 505, 35-41 (1984).
Field J. E., van der Zwaag S. and Townsend D., Proc. 6th Int. Conf. on Erosion by Liquid and Solid impact, Cambridge UK, 21-1 to 21-13 (1983).
Flahaut J., Handbook on the physics and chemistry of rare earths, Ed. Gshneider Jr. K. A. and Eyring L., North Holland 1-87 (1979).
Gentilman R. L., Proc. SPIE 683 2-11 (1986).
Hackworth J. V., Proc. SPIE 362 123-136 (1982).
Hasselman D. P. H., Ceram. Bull. 49, 1033-1037 (1970).
Harris D. C., Hills M. E., Gentilman R. L., Saunders K. J. and Wong T. Y., Advanced Ceramic Materials 2 74-78 (1987).
Klein C. A., diBenedetto B. and Pappis J., Opt. Eng. 25, 519-531 (1986).
Lewis J. C. and Beswick J. A., Allen Clark Research Centre Annual Review 123-129 (1985).
Lewis J. C., Wilson A. E. J. and Beswick J. A., ECOOSA March 22-25 Birmingham UK (1988).

Provenzano P. L., Thesis Crystal Chemistry, Vibrational Spectra and   Luminescence
Studies of Rare Earth Ternary Sulphides Penn. State Univ.   (1976).
    Saunders K. J., Wong T. Y., Hartnett T. M., Tustison R. W. and Gentilman   R. L., Proc.
SPIE 683 72-78 (1986).
    Savage J. A. and Lewis K. L., Proc. SPIE 683 79-84 (1986).
    Walker P. and Ward R. C. C., Mat. Res. Bull. 19, 717-725 (1984).

Table 1. Physical Properties of some 8-12 μm Window Materials

PROPERTY	Ge *	GaAs *	CVD ZnS *	CVD MULTISPECTRAL ZnS *	CaS:La$_2$S$_3$ ratio (X:Y) +	CVD ZnSe *
Refractive Index 10 μm	4.0032	3.2769	2.2002	2.2008	~2.53 (42.5:57.5)	2.4070
Density Kgm^{-3}10^3	5.32	5.32	4.08	4.09	4.26 (42.5:57.5)	5.27
Grain Size μm	cm→mono	cm→mono	2→6	30→120	10→30 (42.5:57.5)	70
Hardness GPa	8.3	7.4	2.2	1.6	5.6 (42.5:57.5)	1.0
Fracture Toughness MNm$^{-3/2}$	0.6 †	–	0.65	1.0	0.53 (50:50)	0.7
Rupture Modulus MPa	100	71	97	69	80 (50:50)	55
Young's Modulus GPa	103	85	75	88	96 (42.5:57.5)	70
Poisson's Ratio	0.279	~0.28	0.29	0.318	0.25 (50:50)	0.28
2mm Water Drop Damage Threshold Velocity m/s	205	–	170	197	250 (42.5:57.5)	137→152
Thermal Expansion Coefficient x 10^{-6}/°C	6.1	5.7	7.4	7.4	14.7 (42.5:57.5)	7.57
Thermal Conductivity W/mK	70	35	17	27	1.7 (42.5:57.5)	18
R'	8.0	3.7	2.1	2.0	0.07	1.4

*   Manufacturers data with the exception of R' which is calculated and water drop damage
    threshold which is taken from Hackworth 1982 and Field et al 1983.

+   Lewis and Beswick (1985) Saunders et al (1986)

†   Blair (1981)

Table 2.

WAVELENGTH	TEMPERATURE	TRANSMITTANCE %					
		Ge	GaAs	ZnS	Multi ZnS	CaS: La$_2$S$_3$ 45:55	ZnSe
8 μm 1250 cm^{-1}	20°C 175°C	47.0 (100°C, 46.2)	55.6 55.1	75.0 74.7	75.0 75.0	68.1 68.2	70.6 70.6
10 μm 1000 cm^{-1}	20°C 175°C	46.8 (100°C, 46.0)	55.7 55.2	75.0 74.4	75.3 75.0	68.7 68.8	70.6 70.6
12 μm 833 cm^{-1}	20°C 175°C	45.2 (100°C, 44.1)	55.1 54.3	73.3 71.1	73.3 71.2	68.2 67.4	70.1 70.1

The in-line transmittance values taken at 3 wavelengths and two temperatures are listed here for all of the 1mm thick samples shown in figure 1.

Figure 1.   In line transmittance of 1mm thick samples of ZnSe, GaAs, Ge, ZnS and CaS45: La_2S_355 at 20°C (solid curve) and 175°C (broken curve) (Ge 100°C).

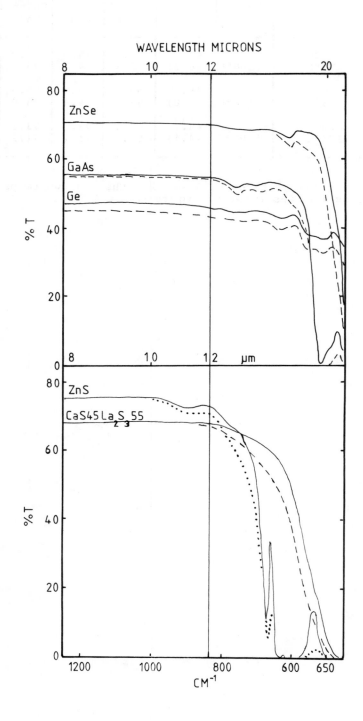

Current and Emerging Materials for LWIR External Windows

R. W. Tustison and R. L. Gentilman
Raytheon Company, Research Division
Lexington, MA 02173-7899

ABSTRACT

Several long wavelength infrared (LWIR) transparent external window materials are compared with respect to their IR bandpass, rain erosion and thermal shock resistance. The materials of interest include: ZnSe, ZnS, multispectral ZnS, calcium lanthanum sulfide (CLS), Ge, GaAs and GaP. Relevant thermal and mechanical property data are presented. The current state of development and size capabilities of each material are also discussed.

1. INTRODUCTION

The search for a high durability, multispectral infrared transparent material has been long and generally unsuccessful. Chemically vapor deposited, polycrystalline ZnS and ZnSe are currently the best available compromise materials, the former being more transparent and the latter being more durable. Ternary sulfides, most notably calcium lanthanum sulfide (CLS) are also under development and would offer further improvement in durability. Of the semiconductors, Ge, GaAs and GaP are either being used or are being considered for some LWIR applications. However, only a limited amount of information exists concerning their optical, mechanical and rain erosion resistance of these latter materials[1]. In this paper we make a comparison of the mechanical, optical and rain erosion data for several candidate LWIR materials.

2. OPTICAL PROPERTIES

A primary requirement for any window material is that it be transparent over the wavelength region of interest. In all optical materials, this range of transparency is established at short wavelengths by the optical band gap of the material and at long wavelengths by atomic vibrational modes. Unfortunately, materials that tend to be strongly bonded and therefore more durable do not typically transmit very far into the infrared. Conversely, weak materials, with small interatomic force constants and large atomic masses often transmit into the far infrared but are not durable.

In practice, extrinsic factors also limit the degree of transparency. These extrinsic limitations are typically associated with specific materials preparation methods. For example, impurity incorporation into the material during processing can cause characteristic vibrational absorptions or these impurities can cause electronic absorptions if they contribute free charge carriers. Particulates or second phase inclusions can also lead to scatter, depending on the size and optical properties of the second phase. The latter is most problematic for optical materials processed from a powder precursor (e.g., CLS).

Figure 1 is a compilation of LWIR transmittance curves for 3 mm thick samples of chemically vapor deposited ZnS and ZnSe as well as for CLS, produced by a powder processing technique.[2] Typical transmittance curves for GaAs, GaP and Ge are also shown for comparison. Figure 2 summarizes the approximate range of transparency for these materials. Only ZnS, ZnSe, CLS and GaP have significant transmittance in the visible through the LWIR.

3. RAIN EROSION DAMAGE

The term rain erosion is really a misnomer when applied to the raindrop impact damage commonly observed in LWIR transparent materials. Little if any material is actually removed before the substrate has suffered substantial damage. This damage is manifested by the formation of cracks produced during

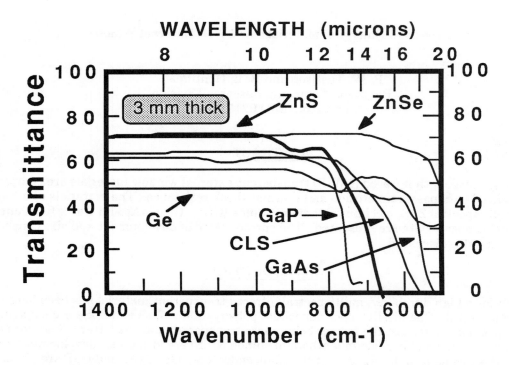

Figure 1.   LWIR transmittance of six window materials (3 mm thick). Standard ZnS and Multispectral ZnS have identical transmittances in this wavelength band.

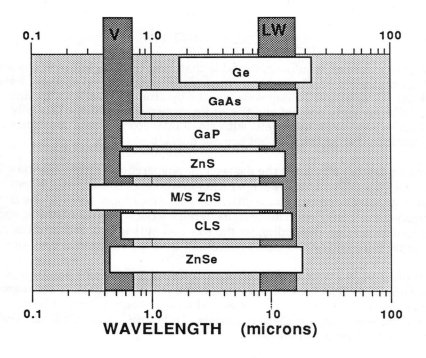

Figure 2.   Range of transmittance of 7 LWIR window materials. In the visible (V) band, ZnSe and Multispectral ZnS are highly transparent; Ge and GaAs are opaque; GaP, CLS and ZnS are somewhat transparent.

the raindrop impact. These cracks are formed during the elastic response of the material to the drop impact. Because of the circular symmetry associated with the raindrop, the cracks typically have an identifiable radius of curvature and are known as circumferential cracks. Collectively, all of the circumferential cracks associated with any one impact event are called a ring fracture. For a complete description of this phenomena the reader is referred to several excellent review articles.[3-6] The onset of rain erosion damage, defined by the velocity at which damage is first observed or the damage threshold velocity, is in the subsonic velocity regime for all known LWIR materials.

Various methods have been suggested in an effort to quantify the onset of rain erosion damage. From an engineering viewpoint, a measurement of the decrease in transmittance of a sample following raindrop exposure would be the most obvious measure of the onset of damage. This decrease might be measured over the relevant bandpass, e.g. 8 - 14 μm. In fact, specifications have been written for some applications that place limits on the amount of transmittance degradation allowed for a fixed exposure time, at a specified velocity. Unfortunately, significant and potentially catastrophic amounts of damage usually occurs before LWIR transmittance degradation becomes measurable.

Van Der Zwaag and Field[4] have suggested that the decrease in fracture strength that accompanies rain erosion damage is a sensitive indicator of the damage threshold in LWIR materials. They define this after-exposure strength as residual strength and have used this to establish the threshold velocity, namely the velocity at which the residual strength decreased below the unexposed value.

The most direct method of determining if damage has occurred is to inspect the sample visually following exposure. Circumferential cracks can be easily identified by their radius of curvature and thereby unambiguously associated with a unique ring fracture. For the sake of comparison, we equate a damage site with the identification of a ring fracture and use this to estimate relative amounts of damage in different materials exposed to the same conditions. We have found that the majority of the observable cracks, particularly at velocities near the threshold velocity, can be associated with a unique damage site. Utilizing the last two methods, namely the determination of residual strength and visual inspection, we have estimated the damage threshold velocity of several LWIR materials.

4. RAIN EROSION TESTING

Samples, 2.22 cm in diameter by 2.54 mm thick were fabricated from material supplied by Raytheon Company, with the exception of Ge, which was supplied by Exotic Materials, Inc. Prior to testing, all of the samples were polished to a surface finish of 60/40 or better.

All of the test results reported herein were collected at the Rotating Arm Facility at Wright-Patterson Air Force Base. The majority of the samples tested were exposed for a period of 5 minutes to a simulated rainfield of 2.54 cm per hour at a 90° impact angle, at several velocities. The drop size was constant at 2.0 ± 0.2 mm diameter.

All exposed samples were examined microscopically. The magnification was limited to 56X so that each sample could be inspected in a reasonable amount of time. Circumferential cracks in excess of 10 microns in length were identified and their associated damage site was counted.

Figure 3 is a micrograph of a typical ring fracture observed in ZnS, using transmitted light. The circumferential cracks have formed near the surface and have propagated into the material, leading to scattering of visible light and hence to the darkened regions in the micrograph. Also note that very little surface pitting or material removal has occurred for this exposure condition. In spite of these cracks, very little corresponding decrease in transmittance has occured in the LWIR.

Figure 4 shows two micrographs of a large grain Ge sample, both uncoated and coated. In this case, substantial material is eroded at relatively low velocities,in the uncoated case. This erosion is also typical of GaAs and GaP. Application of a hard carbon coating[7] clearly reduces the amount of damage and

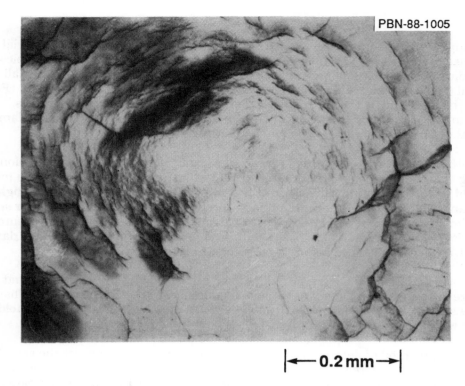

PBN-88-1005

|←— **0.2 mm** —→|

Figure 3.   Transmitted light micrograph of typical ring fracture in ZnS. The sample was tested for 5 min at 201 m/sec in a 2.5 cm/hr rainfield with 2 mm diameter drops.

5 min. @ 210 m/sec   PBN-87-1195

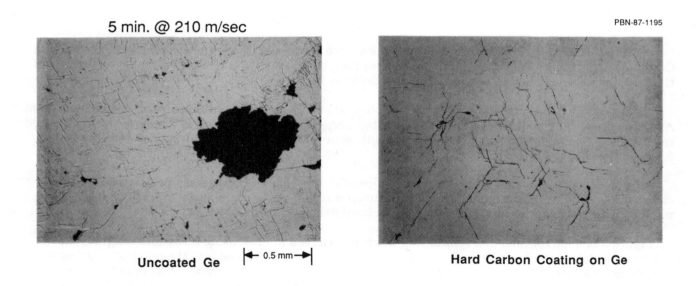

Uncoated Ge |←— 0.5 mm —→| **Hard Carbon Coating on Ge**

Figure 4.   Impact damage in large grain Ge tested for 5 min at 210 cm/hr rainfield with 2 mm diameter drops. Left: Uncoated. Right: With hard carbon coating[7]. The coating offers subtantial improvement. Note the crystallographic nature of the fracturing.

substantially reduces the eroding at this velocity, but damage does occur nevertheless. Also note the crystallographic nature of the fracturing, characteristic of large grain or single crystalline semiconductor material like Ge, Si, GaAs and GaP.

Theory predicts that the damage threshold velocity should depend on two materials parameters,[3] namely the fracture toughness and the elastic wave velocity, c_t = (modulus/density)$^{1/2}$, as well as the size distribution of pre-existent surface cracks. Evans[8] has proposed that a damage parameter can be defined as the product of these two materials parameters, namely $K_c^{2/3} c_t^{1/3}$. Here the contribution of the intial flaw size distribution is not taken into consideration, but might be expected to be similar for materials prepared by the same technique. Previous experimental data on nylon sphere impacts has been shown to be consistent with this analysis.

Accordingly, the single drop threshold velocity data collected by Hackworth and Kocher[9] and by Field et al.[10] is plotted in Figure 5 against this damage parameter. As predicted, the damage threshold velocities for these materials are proportional to this damage parameter. Figure 6 summarizes the velocity threshold data determined for several materials exposed for 5 minutes (about 2000 to 4000 impacts). Again the rough proportionality of the threshold velocity to the damage parameter defined earlier exists. Figure 7 combines both of these sets of data into one plot. Parenthetically, it is clear that the threshold velocity also depends on the number of drop impacts, which is not in agreement with a previous report.[4]

With the validity of the rain erosion parameter established both experimentally and theoretically, it is now possible to arrange the candidate LWIR materials in increasing order of raindrop impact damage resistance, namely: M/S ZnS > GaP > ZnS > ZnSe > Ge > GaAs.

5. THERMAL SHOCK RESISTANCE

For high speed applications, a critical factor for IR windows is resistance to thermal shock from rapid aerodynamic heating. For the more severe flight profiles, maximum tensile stresses typically occur within a few seconds, and if sufficiently severe, cause fracture and catastrophic failure. A specific window material, window geometry, attachment design, and flight profile can be modelled using computer codes to determine the thermally induced stresses and probability of survival. However, such analyses require a complete set of thermal and mechanical properties for the material covering all temperatures of interest, as well as detailed fracture strength statistics.

In lieu of a detailed analysis, thermal shock figures of merit have been established for comparing different materials. The most appropriate figure of merit for resistance to fracture from transient heating is given by:

$$R' = \frac{\sigma (1-\nu) k}{\alpha E}$$

where: σ = fracture strength
ν = Poisson's ratio
k = thermal conductivity
α = thermal expansion coefficient
E = Young's modulus

The thermal and mechanical data and thermal shock resistance parameters at room temperature and at 500°C are given below. The R' values are also shown graphically in Figure 8.

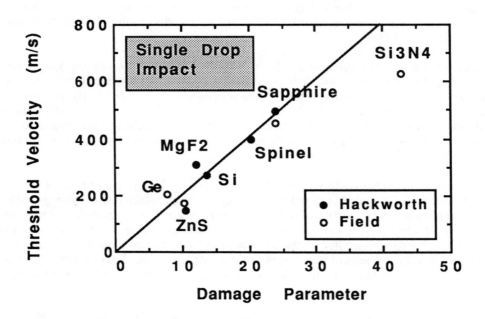

Figure 5.   Single drop rain erosion damage threshold velocities for several materials as a function of damage parameter ($K_c^{2/3} c_t^{1/3}$).

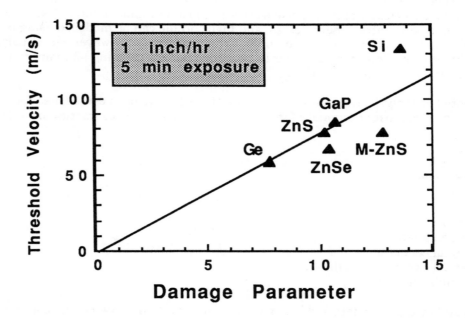

Figure 6.   Multiple drop (2000 to 4000 impacts) rain erosion damage threshold velocities for several materials as a function of damage parameter ($K_c^{2/3} c_t^{1/3}$).

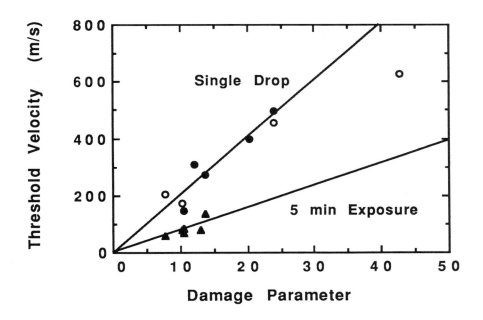

Figure 7.   Combined damage threshold velocity data from Figures 5 and 6. Both data show a linear dependence. However, the threshold velocities for 5 minute exposure are approximately one-half those determined from single drop tests.

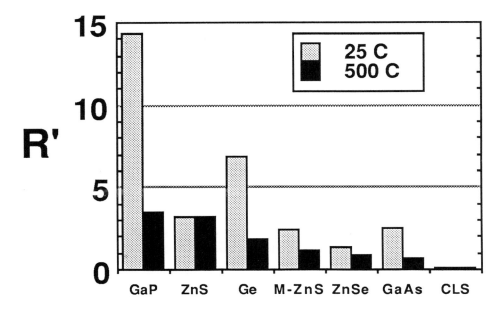

Figure 8.   Thermal shock resistance figures of merit (R') for 7 LWIR window materials at room temperature and 500°C. The materials are ranked based on the 500°C values. Thermal shock resistance decreases with increasing temperature due to a decrease in thermal conductivity. However, compensating strength and Young's modulus changes for ZnS make it equally resistant at both temperatures..

25°C --	Strength MPa	Thermal Conduct. W/m-K	Poisson Ratio	CTE ppm/K	Young's Modulus GPa	R'
ZnSe	52[R]	18[S]	0.28[R]	7.1[R]	70[R]	1.35
ZnS	104[R]	23[S]	0.29[R]	7.0[R]	75[R]	3.22
M/S ZnS	69[R]	32[S]	0.32[R]	7.0[R]	88[R]	2.44
CLS	83[R]	2[S]	0.30[e]	14.7[S]	94[S]	0.08
Ge	93[K]	60[T]	0.28[K]	5.7[T]	103[K]	6.85
GaAs	55[K]	33[K]	0.31[K]	5.7[T]	86[K]	2.56
GaP	104[K]	97[K]	0.31[K]	4.7[T]	103[K]	14.31

500°C --	Strength MPa	Thermal Conduct. W/m-K	Poisson Ratio	CTE ppm/K	Young's Modulus GPa	R'
ZnSe	65[R,S]	7[S]	0.28[R]	7.6[R]	52[R]	0.83
ZnS	152[R,S]	10[S]	0.29[R]	7.0[R]	45[R]	3.24
M/S ZnS	69[e]	10[S]	0.32[R]	7.0[S]	53[R]	1.20
CLS	70[R,S]	2[T]	0.30[e]	14.7[S]	94[e]	0.07
Ge	93[e]	20[T]	0.28[K]	7.0[T]	103[e]	1.86
GaAs	55[e]	8[T]	0.31[K]	7.0[T]	86[e]	0.70
GaP	104[e]	30[e]	0.31[K]	6.0[T]	103[e]	3.47

R = Raytheon measurements
S = Southern Research Institute[11,12]
K = Klocek et al.[13]
T = Touloukian et al.[14]
e = estimated or extrapolated

GaP stands out as the most thermal shock resistant material and CLS as the least resistant. Since Ge becomes conductive and opaque when heated to 200°C, ZnS ranks second with respect to thermal shock resistance. In addition, the thermal shock resistance of ZnS is essentially the same at room temperature and 500°C.

In comparing materials in this way, it is important to note that thermal conductivity and, to a lesser extent, thermal expansion and fracture strength are temperature dependent. Thermal conductivity decreases rapidly with increasing temperature by at least a factor of 2 from room temperature to 500°C. Thermal conductivity is also an extrinsic material property which is highly sensitive to composition, purity, and microstructure.

The fracture strength of a brittle material is a nebulous property -- it depends on sample size, test method, surface finish, and furthermore it is statistical in nature. At least 20 identical samples should be tested for reliable fracture strength data.

When comparing different materials, attention should be given to differences in test procedures. Thermal conductivity and fracture strength data taken from different sources must be evaluated carefully.

6. OXIDATION RESISTANCE OF ZnS WINDOWS

ZnS has been thought to be an unsuitable material at the temperatures expected for very high speed missile flights due to oxidation and/or decomposition. The elevated temperature transmittance of ZnS, ZnSe and CLS was measured recently.[16] This study reported on the shift of the multiphonon absorption band with temperature. A secondary result of the study was the observation that at up to 500°C there was no permanent damage to these materials in air for times exceeding 15 minutes, while surface oxidation occurred at 600°C.

Since the anticipated dome exposure times may be less than 1 minute, short term oxidation experiments with ZnS were conducted recently at Raytheon. Polished disks were exposed to elevated temperatures in stagnant air. A sample was inserted rapidly into a pre-heated furnace, held for a predetermined time (usually 5 minutes), and then rapidly removed. The IR transmittance and weight was measured before and after the exposure. Thermocouple readings indicated that at least 1 minute was required for the sample to reach thermal equilibrium. Thus, for the 5 minute exposures, the samples were at temperature for at least 3 minutes.

The weight loss and 10μm transmittance loss results are shown in Figure 9. Based on these experiments, ZnS would be usable to 800°C for a few minutes.

It should be pointed out that these experiments did not simulate the effects of air flow or pressure during missile flight. Furthermore, the heating rate was not nearly as rapid. More sophisticated experiments will be needed to determine whether ZnS could be used at even higher temperatures for times less than 3 minutes. However, it is clear that ZnS can be seriously considered for short time supersonic missions for dome temperatures up to at least 800°C.

7. DEVELOPMENT STATUS

ZnSe, ZnS, multispectral ZnS and Ge are mature IR window materials, while CLS, GaAs and GaP are in various stages of development. The fabrication process, size availability, and relative cost of the window materials are summarized below.

Mat'l	Fab. Process	Max Size	Status	Relative Cost
ZnSe	Vapor Deposited	100 cm dia.	Commercial	2.5
ZnS	Vapor Deposited	100 cm dia.	Commercial	1.0 Basis
M/S ZnS	Vapor Deposited	50 cm dia.	Commercial	1.3
CLS	Powder	5 cm dia.	Research	4.0 est.
Ge	Melt-Cast	60 cm dia.	Commercial	1.0
GaAs	Bridgman or CVD	30 cm dia.	Development	n/a
GaP	LECz or CVD	5 cm dia.	Research	n/a

8. SUMMARY

Among the current and emerging LWIR materials reviewed herein, each material has certain advantages and disadvantages and no one material meets all needs. The positive and negative attributes are summarized below.

	Advantages	Disadvantages
ZnSe	Full 8-14μm Band Multispectral	Erosion, Cost
ZnS	Erosion, Thermal Shock, Cost	Reduced Bandpass
M/S ZnS	Erosion, Multispectral	Reduced Bandpass
CLS	Erosion, Band Pass	Thermal Shock, Cost
Ge	Full 8-14μm Band, Cost	Opaque when Heated, Erosion
GaAs	Full 8-14μm Band	Erosion
GaP	Erosion, Thermal Shock	Reduced Bandpass

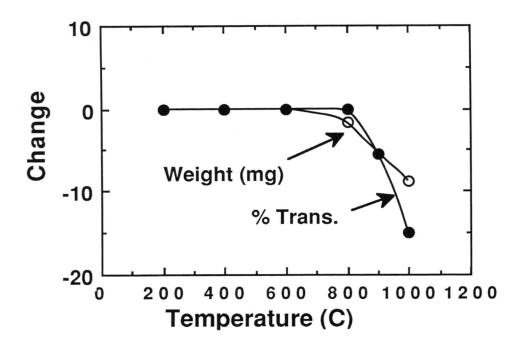

Figure 9.   Weight loss and transmittance loss at 10 microns of ZnS after 5 minute exposure in air at various elevated temperatures. ZnS should be usable to 800°C for a few minutes.

8. REFERENCES

1.   P. Klocek, L. E. Stone, M. W. Boucher and C. DeMilo, Proc. SPIE, 929 , 1988.
2.   R. L. Gentilman, T. Y. Wong, M. B. Dekosky, R. W. Tustison, and M. E. Hills, "Calcium Lanthanum Sulfide as a Long Wavelength IR Material," SPIE Proc. 929, April 1988, 57.
3.   A. G. Evans, Y. M. Ito and M. Rosenblatt, J. Appl. Phys., 51 (1980) 2473.
4.   S. Van Der Zwaag and J. E. Field, J. Mat. Sci., 17 (1982) 2625.
5.   W. F. Adler, J. Mat. Sci., 12, (1977) 1253.
6.   A. G. Evans and T. R. Wilshaw, J. Mat. Sci., 12 (1977) 97.
7.   Hard Carbon coatings were produced by Exotic Materials Inc. of Costa Mesa, CA.
8.   A. G. Evans, J. Appl. Phys. 49, 3304 (1978).
9.   J. V. Hackworth and L. H. Kocher, Final Report AFML-TR-78-184, Bell Aerospace Textron, Buffalo, N.Y.
10. J. E. Field, S. van der Zwaag, D. Townsend and J. P. Dear, Final Report No. AFWAL-TR-83-4101, October, 1983.
11. J. R. Koenig, "Thermostructural Evaluation of Four Infrared Seeker Dome Materials, Part 2. Thermal and Mechanical Properties," NWC TP 6539 Part 2, April 1985.
12. Southern Research Institute, "Thermal and Mechanical Properties of Calcium Lanthanum Sulfide," SoRI-EAS-85-401-5267-I-F, April 1985.
13. Y. S. Touluokian et al. Thermophysical Properties of Matter, Vols. 1,2,12,13, Thermophysical Properties Research Center, Purdue University, Plenum Publishing, 1975.
14. C. M. Freeland, "High Temperature Transmission Measurements of IR Window Materials," SPIE Proc. 929, April 1988, 79.

Phase diagram studies of ZnS systems

J. M. Zhang, W. W. Chen, B. Dunn and A. J. Ardell

Department of Materials Science and Engineering
University of California
Los Angeles, CA 90024

ABSTRACT

The development of infrared-transmitting ZnS-based ceramics with a combination of high strength, hardness and toughness over a range of temperatures is highly desirable. The approach we are currently pursuing is a familiar one to metallurgists, namely the exploitation of thermal, and possibly mechanical, processing of ZnS-base "alloys". Knowledge of the phase equilibria of various ZnS-rich systems is essential to achieve our objectives. Unfortunately, the literature on this subject is sparse, and we have had to undertake such investigations ourselves. This paper describes the results of our initial studies of the solid-state phase equilibria in the ZnS-CdS and ZnS-Ga$_2$S$_3$ phase diagrams. We also discuss possible processing routes to achieve hard and tough ceramics utilizing the phase diagrams established.

1. INTRODUCTION

ZnS has been widely used as an infrared window material because of its good infrared transmission up to 12 μm. However, as a window material, good mechanical properties, especially fracture toughness, thermal shock resistance and erosion resistance are also required. The mechanical properties of single-phase ZnS usually limit its application in many situations. The objective of our research is to enhance the mechanical properties of ZnS, yet retain its good infrared transmission characteristics.

It is well known that the addition of another component to a matrix material can harden it significantly. This can come about simply by solid-solution strengthening or by a composite materials approach if the addition is in the form of a new phase. Alternatively, a combination of thermal and mechanical treatments (e. g. precipitation strengthening) can be employed to introduce and manipulate the dispersion of second phase particles. In any case, the mechanical properties of the material are usually greatly improved. These techniques are already well established for metallic alloy systems and have been used for many years. For instance, precipitation hardening is the principal method of strengthening almost all the wrought aluminum alloys in use today. Transformation toughening of zirconia-based ceramics is another method that exploits the precipitation of particles of a second phase to vastly improve mechanical behavior.

For optical materials it is also necessary to consider the optical properties of the materials after particles of a second phase have been introduced. Care must be taken that the second-phase particles neither absorb nor scatter in the infrared. Absorption losses are controlled by selecting composition systems in which the second phase possesses good infrared transmission. Scattering losses present a more difficult problem. In the present research program, however, the wavelengths of interest are in the range of 10 μm, and the optical losses from scattering can be minimized by keeping the particle size of the second phase much smaller than the wavelength of the incident radiation. Fortunately, small particle sizes are also preferred for improved mechanical properties, often benefitting the strength, fracture toughness and hardness of the ceramic simultaneously. From this perspective, therefore, it is probable that ZnS ceramics can be strengthened by second phase particles, still retaining their good infrared transmission characteristics, if the particle sizes can be controlled such that they are much smaller than the infrared wavelengths involved.

The key to the success of our objective is the choice of a suitable ZnS-based ceramic system, and good candidates must satisfy certain requirements: (1) They should form a solid solution with ZnS at the ZnS-rich end, which will make it possible to solid-solution strengthen or precipitation harden the ceramics; (2) The component material added should be a good infrared-transmitting material itself, otherwise strong absorption might occur; (3) The refractive index of the addition should be close to that of ZnS to avoid substantial scattering losses, although this is not a serious problem if the particles introduced have sizes much smaller than the infrared spectrum.

These objectives can be met, in principle, by manipulation of the composition and thermal treatment of binary ZnS-based ceramic "alloys"; all that is required is an established compilation of binary phase diagrams. Unfortunately, the number of published phase diagrams involving various sulfides, not to mention zinc sulfide, is surprisingly small. In fact, we were initially aware of only two or three ZnS-metal sulfide phase diagrams despite an exhaustive search of the literature, and these were not really suitable for our purposes. We therefore decided to determine the solid-state phase equilibria of several potentially promising candidate binary systems. The ZnS-CdS and ZnS-Ga$_2$S$_3$ systems were chosen for initial study because they most closely satisfy the criteria stated above.

2. THE ZnS-CdS SYSTEM

CdS possesses excellent infrared transmission properties and the system ZnS-CdS offers considerable promise for developing a two phase microstructure. The ZnS-CdS phase diagram has been the subject of several studies. Ballentyne and Ray[1] investigated ZnS-CdS bulk materials annealed at 1100 °C and found solid solution formation over the complete range of compositions. A study of the phase equilibria in thin films in the ZnS-CdS system has also been reported[2]. For films annealed at 450 °C, Kane et al.[2] found the presence of the wurtzite (hexagonal) and sphalerite (cubic) phases at < 60% and > 85% ZnS, respectively. At intermediate compositions a two-phase region consisting of cubic and hexagonal ZnS solid solutions was observed. These authors did not study the phase equilibria as a function of temperature. A more detailed description of the cubic and hexagonal solid solution boundaries was reported by Kaneko et al.[3] This work, however, was conducted under hydrothermal conditions which are not usually relevant for ceramic processing considerations. As of yet, no results have been presented which accurately map the wurtzite/sphalerite phase boundaries under conditions which are amenable to ceramic processing. A knowledge of this phase boundary is essential if one is to select the proper composition and annealing temperature, and to control the amount of second phase precipitation for two-phase strengthening.

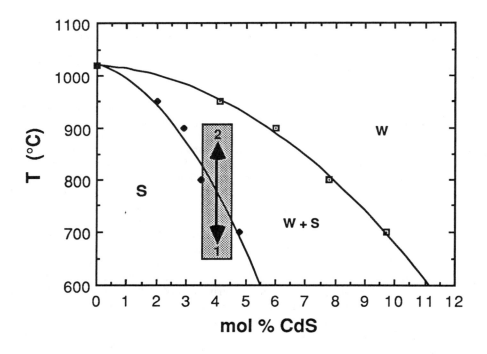

Figure 1. Phase diagram for the ZnS-CdS system in the ZnS-rich region.

The methods we have used to determine the ZnS-CdS phase diagram over the temperature range 700 to 950 °C are described elsewhere[4]. Briefly, the precise compositions of the wurtzite and sphalerite solvuses were determined by lattice parameter measurements and by the disappearing phase method. Both approaches involve the use of X-ray diffraction. A partial phase diagram of the ZnS-rich region is shown in Fig. 1. Some of the more significant features of this diagram are: (1) A range of compositions exists where the sphalerite and wurtzite phases are in equilibrium; (2) The addition of CdS depresses the wurtzite-to-sphalerite transition temperature; (3) The CdS solubility in sphalerite increases as the temperature decreases.

The last finding is particularly noteworthy because alloys which are precipitation strengthened by the classical heat treatment schemes[5] are characterized by having the solute solubility decrease with decreasing temperature. Such behavior is not observed in the ZnS - CdS system, hence a different approach must be utilized for developing the desired two-phase microstructure. One possible route to this end is also depicted in Fig. 1. A dense sample containing ~4 mol % CdS would first be prepared by standard hot-pressing methods[6]. This sample would then be annealed at ~700 °C to form the single phase sphalerite solid solution (point 1 in Fig. 1). A heat treatment at ~900 °C (point 2) would then be made and the wurtzite second phase will nucleate in the cubic solid matrix. Our phase equilibria studies[4] indicate that diffusion in the ZnS matrix is quite slow. Thus, we expect to have considerable control over the nucleation and growth reactions for the wurtzite phase. After heat-treatment, the sample would be rapidly cooled to approximately 600 °C to arrest the precipitation reactions and then slowly cooled to room temperature. Controlled cooling is necessary to prevent thermal shock and, at temperatures below 600 °C, the slow diffusion kinetics ensure that the formation of additional sphalerite solid solution will be minimal.

3. THE ZnS-Ga$_2$S$_3$ SYSTEM

Another series of compositions which display considerable potential for precipitation strengthening with good infrared transmission are in the ZnS-Ga$_2$S$_3$ system. Hahn et al.[7] reported the formation of ZnS solid solutions and the ternary compound, ZnGa$_2$S$_4$ by sintering various compositions of ZnS and Ga$_2$S$_3$. The crystal structure of ZnGa$_2$S$_4$ is tetragonal with lattice constants $a = 0.522$ nm and $c = 1.044$ nm. Single crystals of this ternary chalcogenide have also been grown by an iodine vapor transport method[8]. Phase equilibria studies of the Ga$_2$S$_3$-rich end of the ZnS-Ga$_2$S$_3$ system have been reported[9], but the work which is most relevant for our interests is the ZnS-rich portion investigated by Malevskii[10]. These authors found that there was limited solubility of Ga$_2$S$_3$ in ZnS and that a eutectoid reaction involving the decomposition of wurtzite into sphalerite and ZnGa$_2$S$_4$ occurred. The eutectoid contained 19 mol % Ga$_2$S$_3$ and the eutectoid temperature was 810 °C.

The experimental and analytical methods we have used to establish the ZnS-Ga$_2$S$_3$ phase diagram are similar those employed for the ZnS-CdS system[4]. The compositions prepared for this study varied from 3 to 50 mol % Ga$_2$S$_3$ and the temperatures ranged from 700 to 900 °C. An example of how the parametric method is used to determine the sphalerite solvus is shown in Fig. 2. In the single-phase region the lattice constant varies linearly with Ga$_2$S$_3$ content. The compositions of the sphalerite solvus at 700 and 750 °C are indicated by the intersections of the sloping and horizontal lines.

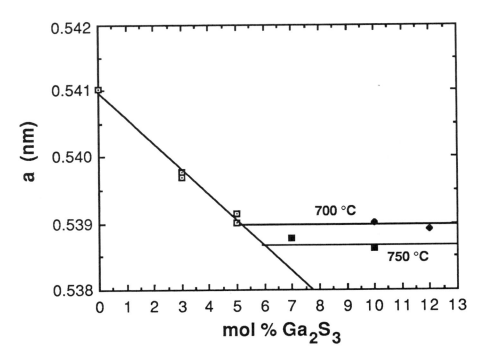

Figure 2. Lattice constant of sphalerite as a function of Ga$_2$S$_3$ content.
The sphalerite solvus at 750 °C is ~6 mol % and that at 700 °C is ~5 mol % Ga$_2$S$_3$.

The phases identified in the present study are superimposed on the phase diagram of Malevskii[10] in Fig. 3 (a). Our initial results indicate that there are discrepancies in the location of the sphalerite solvus and in the composition and temperature of the invariant eutectoid reaction, wurtzite —> sphalerite + tetragonal ($ZnGa_2S_4$). The eutectoid reaction temperature lies between 800 and 850 °C and the eutectoid composition is between 12 and 16 mol % Ga_2S_3. A tentative phase diagram which fits the phase boundaries to our results is shown in Fig. 3 (b). One of the significant features of this phase diagram is that the solute solubility decreases with decreasing temperature (below 800 °C). This behavior is important for achieving precipitation strengthening in metals and alloys[5].

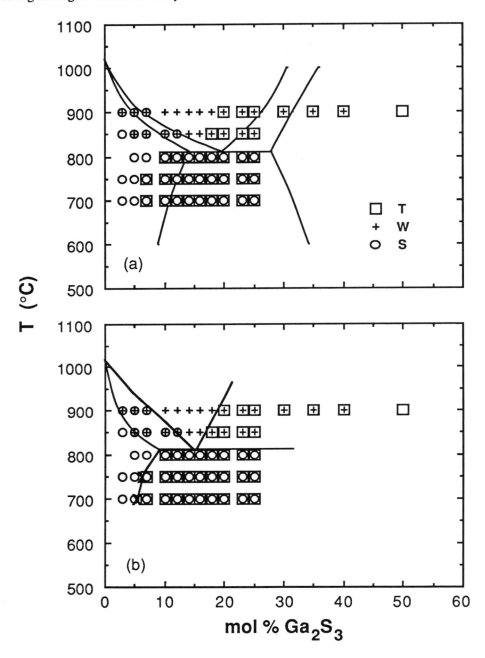

Figure 3. Phase diagrams for the ZnS-Ga_2S_3 system: (a) Data from the present study superimposed on the phase diagram reported by Malevskii[10]; (b) Data from the present investigation. The points at X in (b) represent the values of the solubility limits determined by the intersections of the lines in Fig. 2.

Based on the ZnS-Ga₂S₃ phase diagram, there are several types of heat treatments which should lead to strengthening; these are illustrated in Fig. 4. In at least two cases procedures similar to those employed in metals and alloys can be utilized. The first is analogous to precipitation hardening of alloys[5] (path A in Fig. 4). A dense sample containing ~7 mol % Ga_2S_3 would be prepared by hot pressing and then solution treated at 800 °C to produce a single phase sphalerite solid solution. The nucleation and growth of the tetragonal phase, $ZnGa_2S_4$, could then be induced by heating to the aging temperature of ~650 °C. Some of the fundamental issues to be considered are the time at and magnitudes of both the solution treatment and aging temperatures. These processing variables influence such microstructural features as the grain size of the matrix and the size and concentration of the second phase. In this particular example the precipitation of the tetragonal phase could prove interesting because there is excellent matching of the lattice constants of the tetragonal phase with that of sphalerite (a = 0.541 nm).

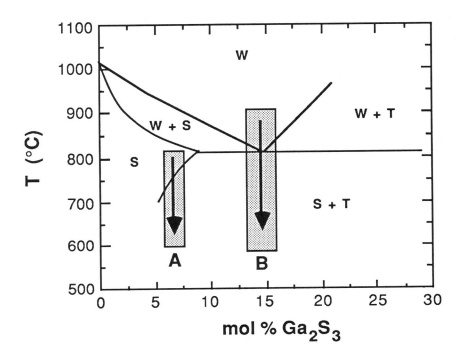

Figure 4. Different heat treatments proposed for strengthening in the ZnS-Ga₂S₃ system.
Path A is analogous to precipitation hardening. Path B involves use of the eutectoid reaction.

The eutectoid reaction (path B in Fig. 4) offers an alternative processing means for introducing a second phase. In this case a solid disc (prepared by hot pressing) of the eutectoid composition (~15 mol % Ga_2S_3) would be solution treated to form the equilibrium wurtzite phase. The sample is then heat treated below the eutectoid temperature, where the hexagonal solid solution is expected to decompose to a mixture of the cubic ZnS and tetragonal $ZnGa_2S_4$ phases. The microstructure here is expected to be inherently different from that in path A, and is likely to be lamellar in morphology, which is typical of eutectoid decomposition reactions[11]. This microstructure can, in turn, be modified by selecting compositions which contain somewhat more or less Ga_2S_3 than the eutectoid composition (i. e., hypoeutectoid or hypereutectoid). It is evident, therefore, that the ZnS-Ga₂S₃ system offers considerable flexibility with respect to the types of processing treatments available for producing second phases in ZnS matrices.

4. CONCLUSIONS

Although the incorporation of a second phase is well known for improving the mechanical properties of ceramics, such multiphase ceramics are generally undesirable for optical transmission, particularly in the visible. In the present program, however, the wavelengths of interest are in the range of 10 μm and control of the precipitation reactions should present little difficulty in keeping the precipitate diameters much less than this dimension.

The initial part of the research has involved phase equilibria studies of suitable ZnS-based systems. The results derived from these studies are essential because they provide basic information regarding the processing treatments required to manipulate and control the ZnS "alloy" microstructures. The solid-state phase equilibria in these two systems are very different. CdS is very soluble in ZnS and the two-phase microstructure envisioned in this system will be composed of sphalerite and wurtzite solid solutions. Ga_2S_3 exhibits limited solubility in ZnS and the ZnS-Ga_2S_3 system is characterized by a eutectoid reaction. Several different types of thermal treatements are possible in this system, ranging from the traditional metallurgical solutionizing anneal, followed by a precipitation reaction, to eutectoid reactions. The diversity of thermal treatments offered by the ZnS-based systems suggests that there will be ample opportunities to develop microstructures which possess enhanced mechanical properties with minimal infrared scattering losses.

5. ACKNOWLEDGEMENT

The authors are grateful to the Office of Naval Research, Contract No. N00014-87-K-0531, for their financial support of this research.

6. REFERENCES

1. D. W. G. Ballentyne and B. Ray, *Electroluminescence and Crystal Structure in the Alloys System ZnS-CdS*, Physica **27**, 337-341 (1961).

2. W. M. Kane, J. P. Spratt, L. W. Hershinger and I. H. Khan, *Structual and Optical Properties of Thin Films of $Zn_xCd_{(1-x)}S$*, J. Electrochem. Soc. **113**, 136-138 (1966).

3. S. Kaneko, H. Aoki, Y. Kawahara and F. Imoto, *Solid Solutions and Phase Transformations in the System ZnS-CdS under Hydrothermal Conditions*, J. Electrochem. Soc. **131**, 1445-1446 (1984).

4. W. W. Chen, J. M. Zhang, A. J. Ardell and B. Dunn, *Solid-State Phase Equilibria in the ZnS-CdS System*, submitted for publication.

5. A. J. Ardell, *Precipitation Hardening*, Metall. Trans. **16A**, 2131-2165 (1985).

6. E. Carnall, Jr., *Hot-Pressing of ZnS and CdTe*, J. Am. Ceram. Soc. **55**, 582-583 (1972).

7. H. Hahn, G. Frank, W. Klingler, A.-D. Störger and G. Störger, *Über ternäre Chalkogenide des Aluminiums, Galliums und Indiums mit Zink, Cacmium und Quecksilber*, Z. Anorg. Allg. Chem. **279**, 241-270 (1955).

8. R. Nitsche, H. U. Bölsterli and M. Lichtensteiger, *Crystal Growth by Chemical Transport Reactions—I Binary, Ternary, and Mixed-Crystal Chalcogenides*, J. Phys. Chem. Solids **21**, 199-205 (1961).

9. A. S. Gates and J. G. Edwards, *Vapor Pressures, Vapor Compositions, and Thermodynamics of the $ZnGa_2S_4$-$ZnGa_8S_{13}$ System by the Simultaneous Knudsen and Dynamic Torsion-Effusion Method*, J. Phys. Chem. **82**, 2789-2797 (1978).

10. A. Yu. Malevskii, *Ranges of Isomorphic Replacement in the System ZnS-Ga_2S_3*, in Eksp. Issled. Obl. Geokhim. Kristallogr. Redk. Elem., Akad. Nauk SSSR, Inst. Mineral., Geokhim. Kristallogr. Redk. Elem., 12-20 (1967).

11. M. P. Puls and J. S. Kirkaldy, *The Pearlite Reaction*, Metall. Trans. **3**, 2777-2796 (1972).

A search for improved 8-12 μm infrared-transmitting materials

Curtis E. Johnson, Terrell A. Vanderah, Christopher G. Bauch, and Daniel C. Harris,

Chemistry Division, Research Department
Naval Weapons Center, China Lake, California 93555

ABSTRACT

Efforts to prepare new infrared-transmitting ceramic materials and to improve the mechanical properties of existing materials are in progress. Work on new materials is concentrating on phosphides, sulfides, and mixed phosphide/sulfides. Zinc sulfide with submicron grain size has been prepared from organometallic precursors with the hope of improving strength or fracture toughness. A survey of the reaction of hydrogen sulfide with organometallic compounds was conducted to evaluate this route to ceramic sulfides.

1. INTRODUCTION

A variety of strong ceramic materials are transparent to infrared light in the mid wave (3-5 μm) region, but there are few materials for the long wave (8-12 μm) region. This dilemma arises because the same factors that contribute to long wave transparency (weak chemical bonding between heavy atoms) give rise to low mechanical strength. For the most demanding applications, ceramic windows are needed that are strong, tough, and have high thermal shock resistance. The latter is favored by high strength, high thermal conductivity, low thermal expansion, and low Young's modulus. Our efforts have been aimed at preparing new infrared-transmitting materials and improving the properties of zinc sulfide, which already enjoys wide use.

Our strategy for selecting infrared window materials is based on the following considerations: Transparency in the long wave infrared region requires that the ceramic material have very low vibrational frequencies. Low vibrational frequencies result from weak chemical bonding between heavy atoms with high coordination numbers (numbers of nearest neighbors). Thermal expansion is minimized by a combination of low cation coordination number and high charges of the ions in the crystal.[1] The requirement for heavy atoms eliminates all oxides and other strong ceramics such as BN and Si_3N_4 which absorb in the long wave infrared region. To further enhance transparency, the material should be an insulator or large-band-gap semiconductor whose constituent elements possess noble gas electronic configurations. Chemical inertness and mechanical strength are enhanced in solids with three-dimensional bonding. The trade off is now apparent: The chemical characteristics that give the best thermomechanical properties degrade the optical properties, since the stronger the chemical bond, the higher the vibrational frequency. Our conclusions are:

(1) To retain suitable thermomechanical properties, low vibrational frequencies should be achieved with heavy atoms rather than weak chemical bonding.

(2) Within the optical constraints, the best thermomechanical properties will be attained by three-dimensional crystal structures having intermediate coordination numbers (5 or 6) and highly charged ions.

2. SCREENING OF KNOWN COMPOUNDS.

CdY_2S_4[2,3] is a scantily characterized thiospinel with an average cation coordination number of 5.3. Attempts to prepare single crystals for optical measurements using iodine transport and KBr flux methods have not been successful to date. We were not able to reproduce the reported crystal growth.[3] Experiments will continue using KBr flux and are presently in progress using $CdCl_2$ as a flux. The thermogravimetric curve for CdY_2S_4 under flowing oxygen is shown in Fig. 1; above 650°C, decomposition to a complex mixture of oxides and sulfates occurs. CdY_2S_4 is at least as stable as ZnS, which oxidizes to ZnO at about 600°C under flowing oxygen.[4] Sufficient quantities of this compound were synthesized to provide several laboratories with samples for initial processing experiments, measurement of thermal expansion (predicted to be similar to that of ZnS), and phase transition profiling. Properties such as hardness, strength, and thermal shock resistance are not yet known for CdY_2S_4.

Binary phosphides tend to be hard and refractory, but also tend to be metallic and chemically reactive.[4] The ternary chalcopyrite-type phosphides $ZnSiP_2$ and $ZnGeP_2$,[5] however, are high band-gap semiconductors, with melting points in excess of 1000°C. They are more than twice as hard as ZnS, are stable in oxygen up to 700°C, and are chemically inert in boiling concentrated HCl.[4,6,7] $ZnSiP_2$ crystals were easily grown using a molten tin flux.[5] Transmission spectra of single crystals confirmed the presence of a strong absorption band at 10 µm arising from the SiP_4 tetrahedra, in agreement with other work;[8,6] therefore, this compound was eliminated as a candidate. The transmission spectrum of a single crystal boule slice of $ZnGeP_2$ (provided by A. Wold of Brown University[6]) was measured and the corresponding GeP_4 absorption band occurred at 13 µm. Spectra collected between 25 and 400°C are shown in Fig. 2; the strong band at 13 µm does not blue-shift into the 8-12 µm window up to 400°C.[7] Preliminary processing experiments have been carried out on $ZnGeP_2$ powder. Densification was satisfactory; however, stability problems were encountered that resulted in metallic products. We attempted to prepare $MgGeP_2$ in order to determine its optical properties; however, our research indicated that this compound does not exist,[9] despite literature reports to the contrary.[5]

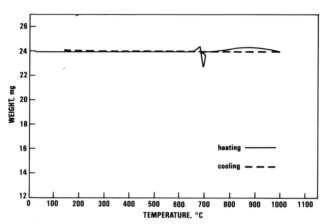

Figure 1. Thermogravimetric analysis of CdY_2S_4 under flowing oxygen.

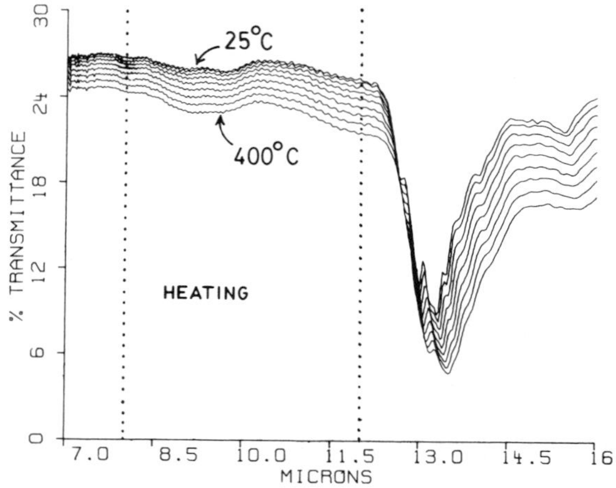

Figure 2. Variable-temperature infrared spectra of single-crystalline $ZnGeP_2$.

3. NEW COMPOUNDS

Approximately 50 new target formulas include phosphides, sulfides, and phosphide/sulfides in various structural families. The formulas include binary analogs to Si_3N_4 (e.g., M_3P_4, M = Si, Ge, Zr), perovskites (e.g., $ABPS_2$, A = Ca, Sr, Ba; B = As, Sb), spinels (e.g., $ZnScZrPS_3$, $Zn[Zr,Hf]_2P_2S_2$, $(Ga,In)[Zr,Hf]_2P_3S$), sodium chloride/nickel arsenide analogs (e.g., M(Sc, Y, La)PS, M = Ca, Sr, Ba), and Si_2N_2O analogs (e.g., Ge_2P_2S).

Synthetic attempts have been carried out for approximately 18 of the new formulas. The binary Si_3N_4 phosphide analogs could not be prepared using solid state methods. All synthetic attempts of ternary or higher formulas have, in general, produced gray, complex mixtures of binary compounds.

Presently, we are focussing our synthetic efforts on solid solutions of CaS and LaP. Both compounds exhibit the rock salt structure and the size disparity between La^{3+} and Ca^{2+} is less than 5%; therefore, the formation of solid solutions seems reasonable. The properties of the two end members are very different, LaP being black and metallic while CaS is a white insulator; the solid solutions may have intermediate properties appropriate for infrared transmission.

[31]P magic angle spinning nuclear magnetic resonance (NMR) spectroscopy is a relatively new characterization tool for inorganic phosphorus-containing solids. We have collected numerous ^{31}P NMR spectra of a variety of materials and have studied the correlation of chemical shift with geometry, formal oxidation state, chemical bond type, and crystallographic site location. The well-resolved spectra of the chalcopyrite-type series $ZnSiP_2$, $ZnGeP_2$, and $ZnSnP_2$ are given in Fig. 3. The trend in chemical shift values for the series does not correlate with chemical periodicity, and this was attributed to an anomalous electronegativity value for germanium.[10] In a study of the solid state ^{31}P NMR spectra of ZrP, Mg_3P_2, MgP_4, and $CdPS_3$, the observed resonance could be assigned to the crystallographically distinct types of phosphorus by considering the structural details of each system.[11]

4. FINE-GRAINED ZINC SULFIDE FROM ORGANOMETALLICS

ZnS is the baseline against which other strong, high temperature, long wave infrared-transmitting materials must be measured. It may be possible to improve the mechanical properties of ZnS, thereby allowing more demanding applications of

ZnS. Currently available commercial material has a grain size of 2-5 μm. It is thought that reducing the grain size by a factor of 10 may enhance mechanical behavior.

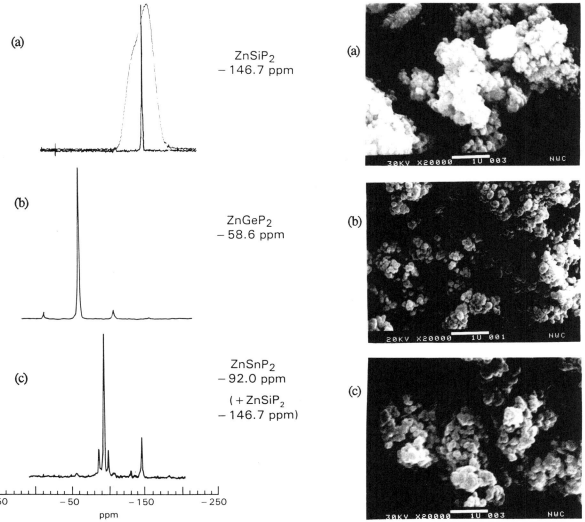

(a) ZnSiP$_2$ − 146.7 ppm

(b) ZnGeP$_2$ − 58.6 ppm

(c) ZnSnP$_2$ − 92.0 ppm

(+ ZnSiP$_2$ − 146.7 ppm)

+ 50   − 50   − 150   − 250
ppm

Figure 3. ^{31}P magic angle spinning NMR spectra of the chalcopyrite-type series (a) ZnSiP$_2$, (b) ZnGeP$_2$, and (c) ZnSnP$_2$.

Figure 4. Electron micrographs of ZnS powder (a) dried at 200°C under vacuum, (b) treated according to Eq. 2, and (c) treated according to Eq. 3.

Several years ago we discovered that ZnS produced by the reaction of H$_2$S with dialkylzinc compounds in organic solvents or the gas phase is a poorly crystalline material consisting of spherical particles generally less than 0.1 μm in diameter.[12,13]

$$(C_2H_5)_2Zn + H_2S \longrightarrow ZnS + 2 C_2H_6 \qquad (1)$$

The fine particles contain carbon impurities at levels between a few tenths of a percent up to several percent. Residual ethyl groups bound to zinc could be reduced to the parts per million level by careful control of reaction conditions, but other carbon-containing impurities were still present at the percent level. When such ZnS powder was compacted to full density by hot pressing or hot isostatic pressing, a black ceramic body resulted, containing grains of 0.1 μm diameter. The Knoop hardness of the black ceramic was almost twice as great as that of commercial optical quality ZnS. The increased hardness and retention of small grain size were encouraging, so we decided to try to remove the carbon impurities to obtain optical quality ZnS.

It was known that ZnS powder could be heated under O_2 to oxidize organic impurities, but literature conditions[14] led to grain sizes up to a micron. A new effort was undertaken to produce carbon-free ZnS powder of submicron grain size. Two procedures were found, one involving gentle oxidation with ozone (Eq. 2) and the other oxidation with oxygen (Eq. 3):

$$\text{ZnS (containing carbon)} \xrightarrow[200°C]{\text{vacuum}} \xrightarrow[25°C]{O_3} \xrightarrow[800°C]{\text{vacuum}} \text{ZnS (carbon free)} \qquad (2)$$

$$\text{ZnS (containing carbon)} \xrightarrow[300°C]{\text{vacuum}} \xrightarrow[400°C]{O_2} \xrightarrow[800°C]{\text{vacuum}} \text{ZnS (carbon free)} \qquad (3)$$

Great care was necessary to prevent runaway oxidation leading to ZnO in the ozone procedure. The reaction is carried out by passing short bursts of dilute O_3 in O_2 through ZnS powder dispersed on a glass frit. For the oxygen heat treatment the ZnS powder is placed in a quartz boat in a tube furnace and dry oxygen is passed over the solid at 400°C. The final necessary step for both treatments is slow heating to 800°C under a vacuum $<10^{-3}$ torr.

Figure 4 shows scanning electron micrographs of ZnS powder prior to oxidative treatment and the powder after undergoing each cleanup procedure. Note that the particle size is essentially unaffected by the treatments although some necking seems evident in the oxygen-treated sample. Figures 5 and 6 show the diffuse reflectance Fourier transform infrared spectra of the ZnS powder at each stage of treatment. The spectrum of the initially dried powder has impurity bands near 3400 and 1600 cm^{-1} assigned to adsorbed water, 2500 cm^{-1} assigned to Zn-S-H groups, and very weak features near 2900 cm^{-1} assigned to organic impurities. (The adsorbed water presumably comes from handling the powder in air to obtain the infrared spectrum. The powder is otherwise synthesized and maintained in a strictly anhydrous atmosphere.) The major effect of the oxidation treatments is the appearance of a peak at 1120 cm^{-1} assigned to $ZnSO_4$. This peak disappears during the final 800°C vacuum heating, during which $ZnSO_4$ should decompose to ZnO. A barely detectable X-ray powder diffraction line for ZnO is observed in the final product. The final powder is much less hygroscopic than the starting powder, presumably from a reduction of surface area.

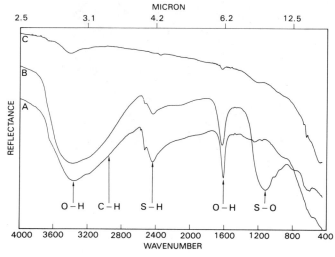

Figure 5. Diffuse reflectance FTIR spectra of ZnS powder (a) dried at 200°C under vacuum, (b) then treated with O_3/O_2 at 25°C, (c) then heated at 800°C under vacuum.

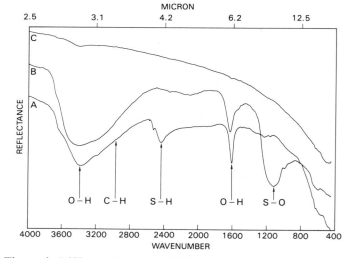

Figure 6. Diffuse reflectance FTIR spectra of ZnS powder (a) dried at 300°C under vacuum, (b) then treated with O_2 at 400°C, (c) then heated at 800°C under vacuum.

Powders treated by the methods described above have been fabricated into ceramics whose mechanical properties will be evaluated. Powder treated according to Eq. 2 gave a yellow ceramic, while powder from Eq. 3 gave a gray ceramic. If the yellow ceramic has improved strength or fracture toughness, the organometallic route to ZnS may become important for the most demanding long wave dome applications.

5. SURVEY OF THE REACTION OF H₂S WITH ORGANOMETALLICS

Among the metals studied, alkyl complexes of Zn, Cd, Hg, and Mg react most completely. Bis(trimethylsilylmethyl)-cadmium gives CdS immediately upon exposure of a toluene solution to H_2S. The product contains 0.2% carbon (by elemental analysis) and acid hydrolysis liberates 4×10^{-5} mol tetramethylsilane per mol of sulfide. Bis(trimethylsilyl-methyl)mercury is much less reactive, and requires exposure of the neat liquid alkyl at 150°C to H_2S to produce black crystalline cubic HgS, which is not the thermodynamically stable phase of HgS at 150°C. Diethylmagnesium is converted to the sulfide, which gives 3×10^{-5} mol ethane per mol of sulfide upon hydrolysis.

Reaction of R_3Al (R = methyl, ethyl, i-butyl) with H_2S gives moisture sensitive white solids containing 0.2-0.6 alkyl groups per Al. Trimethylgallium yields an air-stable white solid product of stoichiometry CH_3GaS, with a CH_3-Ga bond stable to hydrolysis in concentrated H_2SO_4. The product partially sublimes at 390°C and decomposes at 450°C at 10^{-4} torr. Tris(cyclopentadienyl)yttrium in tetrahydrofuran solvent reacts sluggishly with H_2S at 60°C to produce a solid containing considerable solvent. This reaction is still under study.

6. ACKNOWLEDGMENTS

The authors gratefully acknowledge the assistance of C. K. Lowe-Ma, D. K. Hickey, R. W. Woolever, J. H. Johnson, R. A. Nissan, M. P. Nadler and D. L. Decker in many technical aspects of this work. This work was supported by the Office of Naval Research.

7. REFERENCES

1. R.M. Hazen and L.W. Finger, Comparative Crystal Chemistry, p. 136, John Wiley & Sons, New York (1982).
2. O. Schevciw and W.B. White, Mater. Res. Bull. 18, 1059 (1983).
3. A. Tomas, M. Guittard, J. Flahaut, Acta Crystallogr. B42, 364 (1986).
4. J. Covino, Proc. SPIE 505, (1984).
5. J. L. Shay and J.H. Wernick, "Ternary Chalcopyrite Semiconductors: Growth, Electronic Properties, and Applications," Pergamon Press, New York (1975).
6. G.-Q. Yao, H.-S. Shen, R. Kershaw, K. Dwight, and A. Wold, Mater. Res. Bull. 21, 653 (1986).
7. H.-S. Shen, G.-Q. Yao, R. Kershaw, K. Dwight, and A. Wold, Proc. SPIE Meeting, August 1986.
8. G.C. Bhar and R.C. Smith, Phys. Stat. Sol. A13, 157 (1972).
9. T.A. Hewston, J. Solid State Chem. 69, 179 (1987).
10. T.A. Vanderah and R.A. Nissan, J. Phys. Chem. Solids, in press.
11. R.A. Nissan and T.A. Vanderah, J. Phys. Chem. Solids, submitted.
12. C.E. Johnson, D.K. Hickey, and D.C. Harris, Proc. SPIE, 683, 112 (1986).
13. D.C. Harris, R.W. Schwartz, and C. E. Johnson, U.S. Statutory Invention Registration H429 (1988).
14. E. Carnall and L.S. Ladd, U.S. Patent No. 3,131,026 (1964).

SESSION 2

Electro-Optics and Nonlinear Conversion I

Chair
Saluru B. Krupanidhi
Pennsylvania State University

Recent Works on Optical Crystals
in Shanghai Institute of Ceramics, Academia Sinica

Hao-Ran Tan   Chong-Fan He

Shanghai Institute of Ceramics, Academia Sinica
865 Changning Road, Shanghai 200050, China

ABSTRACT

Recent works on bismuth germanate and lithium niobate are described with relation to the behavior of impurity or dopant ions and their effect on optical properties. For bismuth germanate crystals grown by Bridgman method the mean segregation coefficients for several elements were found to be very small as compared with those for Czochralski method. Radiation damage experiments revealed that Pb, Mn, Fe and Co have the most deteriorative effect. Whereas Ni, Ga and Mg affect slightly, and Al, Si, Ca and Cu seem to have no measurable effect on radiation damage of bismuth germanate crystals. For lithium niobate absorption edge of congruent crystals doped with MgO shows a minimum at some 5 mol% Mg. This threshold effect of Mg dopant finds its origin in its substitution behavior in host lattice and could be closely related to the occupancy (4.58%) of Nb in Li site in congruent crystal. For the first time the crystal structure of lithium niobate heavily doped with MgO was investigated directly, resulting in a best model with all the Mg accommodated in Li sublattice. On the basis of experimental data the substitution behavior of Mg could be summarized as the following: With Mg content below the threshold value, replacement of Nb in Li site by Mg is the dominant process, leading to a shift of absorption edge toward shorter wavelengths. As Mg content reaches the threshold value, most of the Nb in Li site are replaced, resulting in the observed minimum of the absorption edge curve. With further increase of Mg content, the substitution of Li by Mg becomes prevailing, causing a slight counter-shift of the absorption edge.

1. INTRODUCTION

Crystal materials, especially electrooptic and optoelectronic crystals, have long been one of the main fields of research in the Shanghai Institute of Ceramics, Academia Sinica. In recent years, research works have been carried out on lithium niobate, lithium tantalate, bismuth germanate, bismuth silicate, cubic zirconia and several kinds of stoichiometric laser crystals. In the present paper experimental results of crystal growth, structure and optical properties with relation to the behavior of impurity or dopant ions and their effect on property for bismuth germanate and lithium niobate are described.

2. BISMUTH GERMANATE

Bismuth germanate (BGO), $Bi_4Ge_3O_{12}$, as a new scintillator material with high atomic number, fast response to radiation, high energy resolution, short fluorescent decay time and nonhydroscopic property, has found its wide application in nuclear physics and high energy physics. Crystals with size up to 3.5 x 3.5 x 27 cm have been grown routinly in the Shanghai Institute of Ceramics to meet the requirement of CERN (European Organization for Nuclear Research) for use in high energy physics. Up to now, 7840 pieces of such large BGO crystal rods have been supplied to CERN LEP-3 Project for construction of an electromagnetic calorimeter. These crystals were grown by Bridgman method with Pt crucibles. For growth of large BGO crystals with good optical quality and reasonable yield, study has been carried out on segregation of impurities, purification of raw materials, stoichiometry, growth parameters, corrosion on Pt crucible, mechanical processing, optical property and radiation damage of BGO crystals. Here some interesting experimental results relevant to the segregation of impurities and their effect on optical property are briefly described[1-4].

2.1 Segregation of impurities

Segregation or the distribution of a certain impurity between melt and growing crystal is a measure of the ability for this impurity to be incorporated in the crystal from melt and can be described in terms of the mean segregation coefficient k, defined as the ratio of the mean concentrations in the bulk crystal, C_s, to the initial concentration of impurity in the melt, C_m. In our experiments, these concentrations were determined by neutron activation analysis. k for several elements with concentration level in the order 0.2 mmol per kg BGO (5-50 ppm by weight) were determined as follows:

Al	< 0.056	Mn	0.022	Fe	0.023
Cr	0.021	Co	0.005	Cu	<0.04

The k values are obviously quite small as compared with those for BGO crystals grown by Czochralski method[5] or for the same elements in other complex oxide crystals. This phenomenon could be reasonably explained in terms of the distinct differences in ionic radii between these impurity ions and the matrix ions Bi or Ge and in the growth conditions.

The small values of segregation coefficients of impurity ions in BGO crystals offer the possibility of purification of BGO crystals by means of multiple crystallization. With this method highly purified single crystal samples have been obtained.

2.2 Effect of impurities on radiation damage

Impurities in BGO crystals have important bearings on their colour and resistance to radiation damage. Experiments for elucidation of the effect of different kinds of impurities in BGO crystals have been carried out on crystal samples synthesized from 5N Bi_2O_3 and 6N GeO_2, repeatedly crystallized three times for purification and doped separately with different metallic oxides at the same doping level 0.2 mmol per kg BGO. Radiation damage was measured by the decrease in relative light output after irradiated with 0.66MeV Cs[137]. Experimental result is summarized in Figure 1, which shows the relationship between the recovery of pulse heights and recovery times for BGO crystals undoped (Samples L-3 and 0-5) and doped with metallic oxides, among which Fe and Pb are also presented in different concentrations (Fe-1 0.10, Fe-2 0.20 and Fe-3 0.30 mmol Fe per kilogram BGO; Pb-1 0.10, Pb-2, 0.20 and Pb-3 0.30 mmol Pb per kilogram BGO). It can be seen from this figure that Pb, Mn, Fe and Co have serious deteriorative effect, the higher their concentrations, the more serious is the effect. Ni, Ga and Mg affect slightly, while Al, Si, Ca and Cu seem to have no measurable effect on radiation damage of BGO crystals. In addition, crystal sample with Ge content higher than stoichiometry (Sample Ge) is more susceptible to radiation damage as compared with the sample with excess Bi (Sample Bi).

It is interesting to note that the absorption spectra for these crystal samples, both doped and undoped, can be analysed as composed of the same three Guassian components correspondent to three absorption centers at about 2.3, 3.0 and 3.8 eV respectively but different in their amplitudes. Based on experimental data a mechanism for radiation damage of BGO taking both intrinsic vacancy and effect of impurity into account was proposed.

3. LITHIUM NIOBATE

Lithium niobate, $LiNbO_3$, having been widely applied in nonlinear optics, electrooptics, piezoelectrics and surface acoustics as a versatile crystal material, is now preparing to play a leading role on the new stage of integrated optics in the near future. An optimization analysis of device-system level has shown that for uses in guided-wave optics lithium niobate appears to be the most desirable material commercially available in good optical quality.[6] It is expected that in U. S. lithium niobate integrated optical devices will grow very rapidly over the decade of the '90s, reaching about $900 million per year by the turn of the century, but as a major barrier the potential for optical damage to this material needs to be better understood and reduced.[7]

Some years ago lithium niobate doped with certain amount of MgO has been found for the first time to have remarkably enhanced resistance to optical damage.[8] The effect of MgO doping was later confirmed,[9, 10] and has aroused widespread interest in lithium niobate research. In this regard a series of papers have been published on crystal growth,[11-13] photoconductivity,[14] ESR,[15] photovoltaic effect,[16] Raman spectra,[17] anisotropic self-diffraction,[18] absorption spectra,[19, 20] electrooptic properties[21] and Ti-indiffused waveguides[22] of MgO doped lithium niobate crystals. But only a few of them studied directly on the substitution behavior of Mg in host crystal lattice. It is believed that a careful and direct investigation upon this fundamental aspect is crucial for a thorough understanding of the mechanism relevant to the effect of Mg ions on photorefractive property in lithium niobate. Some recent experimental results on absorption edge measurement and crystal structure determination of MgO doped lithium niobate are presented here, and the substitution behavior of Mg ions is discussed.[23, 24, 25]

3.1 Crystal growth

Lithium niobate crystals doped with different contents of MgO were grown along the Z-axis by the Czochralski technique from congruent melts containing 2.0, 4.8, 5.7 and 8.3 mol% MgO separately. The ratio of mol% of Li_2O/Nb_2O_5 in congruent melts is 48.6/51.4.

3.2 Absorption edge measurement

UV absorption spectra were measured on c plates cut from the top portions of lithium niobate crystals undoped and doped with MgO by a spectrophotometer in region 300 - 350 nm with a resolution of ±0.1 nm. Absorption data were corrected for multiple reflection on sample surfaces and convered into absorption coefficients. Figure 2 shows the absorption

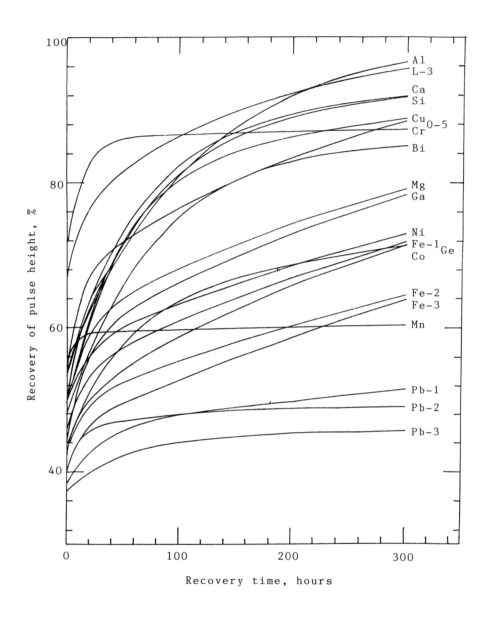

Figure 1. Recovery of pulse height as a function of recovery time in undoped and doped bismuth germanate irradiated with 0.66 MeV Cs137.

coefficients with various wavelengths. As the absorption edges can hardly be determined accurately from these curves, the wavelengths at which the absorption coefficients equal to 20 cm^{-1} are taken as the absorption edges. Figure 3 shows the variation of absorption edges with MgO contents. With increasing MgO content in melts, the absorption edge shifts toward shorter wavelengths at first and then arrives a minimum when MgO content is about 5 mol%. Further increase of MgO content leads to a slow shift of absorption edge toward the longer wavelength direction.

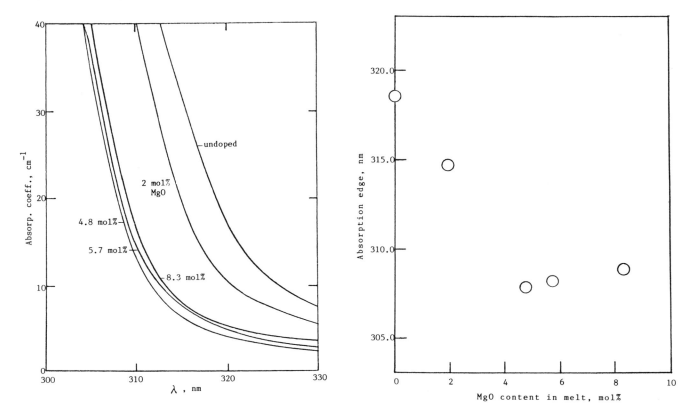

Figure 2. UV absorption coefficients of undoped and MgO doped congruent lithium niobate.

Figure 3. Absorption edges at different MgO contents in congruent lithium niobate.

It has been pointed out by various authors that in MgO doped lithium niobate a series of properties, such as resistance to optical damage, photoconductivity, ESR data, holographic grating erasure time, diffraction efficient, phase-matching temperature and Curie temperature, have been found to be dependent of MgO content and all show a threshold effect at MgO content about 5 mol% . Now this threshold effect also appears in absorption curves with the same threshold content of MgO. This coincidence suggests that there must be certain inherent connection among the observed phenomina. From the point of view of defect chemistry they may have the same origin in the substitution behavior of Mg ions in lithium niobate crystal lattice.

The defect chemistry of lithium niobate has been studied extensively. The cation substitution model for congruent composition proposed by Nassau and Lines[26] has been widely accepted and was exhaustively refined by Abrahams on the basis of X-ray defect structure analysis.[27] This model adopts a filled Li sublattice partially occupied by Nb atoms and creating vacancies in Nb sites. The occupancy of misplaced Nb in Li site and vacancy concentration in Nb site are calculated to be 4.58 and 3.66% respectively for congruent composition.[28] As shown by the crystal structure the misplaced Nb seems to be energetically unstable and would be preferably substituted by Mg in case of MgO doping.

The absorption edge of undoped lithium niobate was interpreted to be correspondent to the fundamental optical transition of an electron from the oxygen p state to the d state of a Nb ion. The misplaced Nb ion in Li site can be treated as impurity ion, which could cause an increase of absorption edge.[19] Thus the replacement of the misplaced Nb by Mg could produce the observed shift of absorption edge toward shorter wavelengths. As the MgO content increases, this substitution proceeds until most of the misplaced Nb are replaced.

In this way the concentration of Nb in Li site can be regarded to be closely related to the threshold value of MgO content which produces the maximum shift of absorption edge toward shorter wavelengths.

It should be pointed out that in MgO doped lithium niobate the replacement of Li ions by Mg ions can not be ruled out, as they have similar ionic radii and less disparity in ionic charge as compared with Nb ion. However, as revealed by absorption measurement on lithium niobate[19, 20] loss of Li ions results in a shift of absorption edge toward longer wavelengths, therefore the replacement of Li by Mg could not be the dominative process in low MgO lithium niobate (i.e. below some 5 mol%, the threshold value). On the other hand as judged by our absorption data, this substitution process would become prevailing after the misplaced Nb ions in Li sites have been replaced at high MgO content (above the threshold value). This substitution behavior of Mg in lithium niobate could give explaination to the shifts in counter directions of absorption edge at MgO contents below and above its threshold value. The following crystal structure determination of high MgO lithium niobate is in accord with the above inference.

3.3 Crystal structure determination

In order to obtain direct evidence of the defect structure of MgO doped lithium niobate, crystal structure of a sample doped with 8.8 mol% MgO in crystal was analysed on a CAD-4 X-ray diffractometer with Mok$_\alpha$ radiation. A group of 25 reflections with $13 \leq \theta \leq 21°$ was measured to obtain the lattice constants. For structure measurement the θ, 2θ step-scan technique with $\Delta 2\theta = 0.85° + 0.35\tan\theta$ was used to collect intensity data. A total of 358 reflections including 290 observable reflections for use in structure refinement with magnitudes larger than 3σ was measured. The structure parameters of Abrahams[27] for congruent lithium niobate were taken as initial values in refinement of MgO doped crystal. Several models for Mg substitution were assigned and refined by the method of least squares under the control of a PDP-11 computer. A preliminary result of site occupancy is shown in Table 1. For comparison the corresponding values for undoped congruent lithium niobate by Abrahams and Marsh[27] are also listed. Further refinement is still under way.

Table 1. Site Occupancy in % Atom for Undoped and MgO Doped LiNbO$_3$

Site	MgO Doped		Undoped[27]	
Li	91.0	Li	94.1(3)	Li
	0	Nb	5.9(3)	Nb
	9.0	Mg		
Nb	0	Li	0	Li
	96.3	Nb	95.3	Nb
O	100	O	100	O

From Table 1 it is notable that in the lithium niobate sample doped with 8.8 mol% MgO all the misplaced Nb in the Li site and 4.1% atom of Li are substituted by Mg, resulting in a total of 9.0% atom Mg occupying the Li site. At the same time the occupancy of Nb in the Nb site slightly increases, leaving less vacancies in the Nb lattice. No Mg was found occupying the Nb site.

In conclusion as indicated by our experimental results of absorption measurement and crystal structure determination for MgO doped lithium niobate, the behavior of substitution of Mg in lithium niobate could be summarized as following. With Mg content lower than the threshold value (some 5 mol%), replacement of Nb in Li site by Mg is the dominant process, leading to a shift of absorption edge toward shorter wavelengths. As Mg content increases to the threshold value, Nb in Li site has a minimum number, resulting in the observed minimum of the absorption edge curve. Further increase of Mg content results in the replacement of Li by Mg, causing a slight shift of the absorption edge toward longer wavelengths.

Detailed result of crystal structure determination for MgO doped lithium niobate including lattice parameters, atomic coordinates, bond distances and angles as well as occupancy will be reported and discussed elsewhere.[23, 25]

4. CONCLUSIONS

Radiation damage in BGO crystals was found to be closely related to the kind and concentration of impurities, among which Pb, Mn, Fe and Co were shown to be most deteriorative. The quite small values of segregation coefficients of impurities offer the possibility of purification of BGO crystal by means of multiple crystallization.

The threshold effect of MgO dopant in lithium niobate crystals was also found in their absorption characteristics. Absorption edge measurement in conjunction with defect structure determination suggests that misplaced Nb in Li site would be preferentially replaced by Mg. As MgO content exceeds the threshold value, replacement of Li in Li site by Mg would become prevailing.

5. ACKNOWLEDGEMENTS

The authors would like to thank their colleagues in the BGO and Lithium Niobate Groups of the Shanghai Institute of Ceramics for their cooperation in performing the reported research works. Valuable help of Professors R. Y. Zhu, H. Stone and H. Newman, California Institute of Technology, Pasadena, U. S. A., in neutron activation analysis and study of radiation damage of BGO crystals are gratefully acknowledged. The authors wish to express their gratitude to Mr. Guang Wu and Professor Min-Qin Chen, Center of Analysis and Measurement, Fudan University, Shanghai, China, for their intimate cooperation and valuable discussion in crystal structure determination of MgO doped lithium niobate. The research work on lithium niobate wave-guide material described in the present paper was supported by the National Natural Science Foundation of China.

6. REFERENCES

1. He Chongfan, Fan Shiji, Liao Jingying, Shen Quan shun, Shen Dingzhong, Zhou Tianqun, Prog. Crystal Growth and Charact. 11 (1985)253.
2. He Chongfan, Fan Shiji, Liao Jingying, Shen Quanshun, Shen Dingzhong, Zhou Tianqun, 7th Chinese Conf. on Crystal Growth (Abstracts), Oct. 1985, Yantai, China (1985) 15
3. Zhou Tianqun, Radiation Damage of Bismuth Germanate Scintillator, Master Thesis, Shanghai Institute of Ceramics (1986).
4. T. Q. Zhou, H. R. Tan, C. F. He, R. Y. Zhu, H. B. Newman, Nucl. Instru. Methods Phys. Res. A258 (1987) 58.
5. R. Barnes, J. Cryst. Growth 69 (1984) 248.
6. R. L. Holman, L. M. Athouse Johnson, D. P. Skinner, Opt. Eng. 26 (1987) 134.
7. S. L. Blum, S. H. Kalos, J. B. Wachtman, Ceramic Industry 125 (1985) 40.
8. Gi-Guo Zhong, Jin Jian, Zhong-Kang Wu, 11th Intern. Quant. Electron. Conf., IEEE Cat. No. 80, CH 1561-0 (1980) 631.
9. D. A. Bryan, R. Gerson, H. E. Tomaschke, Appl. Phys. Lett. 44 (1984) 847.
10. D. A. Bryan, R. R. Rice, R. Gerson, H. E, Tomaschke, K. L. Sweeney, L. E. Halliburton, Opt. Eng. 24 (1985) 138.
11. B. C. Grabmaier, F. Otto, Proc. 8th Intern. Conf. Cryst. Growth, York, UK, July 1986 (1986) 682.
12. B. C. Grabmaier, F. Otto, SPIE 651 (Integrated Opt. Circuit Eng. III) (1986) 2.
13. Zhong Jiguo, Lu Yucai, Lu Changqing, Chen Jiarong, Zou Fuqing, J. Chinese Silicate Soc. 12(1984) 145.
14. R. Gerson, J. F. Kirchhoff, L. E. Halliburton, D. A. Bryan, SPIE 704 (Integrated Opt. Circuit Eng. IV) (1986) 221.
15. K. L. Sweeney, L. E. Halliburton, D. A. Bryan, R. R. Rice, R. Gerson, H. E. Tomaschke, J. Appl. Phys. 57 (1985) 1036.
16. Wang Huafu, Shi Guotang, Wu Zhongkang, Phys. stat. sol.(a) 89 (1985) K211.
17. Liu Simin, Acta Physica Sinica 32 (1983) 103.
18. L. Arizmendi, R. C. Powell, J. Appl. Phys. 61 (1987) 2128.
19. I. Foldvari, K. Plogar, A. Mecseki, Acta Physica Hungarica 55 (1984) 321.
20. Feng Xigi, Liu Jiancheng, Xue Liangying, Wang Zhicheng, Proc. Intern. Symp. Appl. Ferroelectrics (ISAF'86), 1986 (1986) 29.
21. R. J. Holmes, Y. S. Kim, C. D. Brandle, D. M. Smyth, Ferroelectrics 51 (1983) 41.
22. C. H. Bulmer, Electron. Lett. 20 (1984) 902.
23. Ma Yixian, Zhu Quanbao, Wu Yao'an, Tan Haoran, to be published.
24. Ma Yixian, Constituent Segregation and Substitution Behavior of MgO in Lithium Niobate, Master Thesis, Shanghai Inst. of Ceramics (1988).
25. Guang Wu, Yi-Xian Ma, Min-Qin Chen, Hao-Ran Tan, to be published.
26. K. Nassau, M.E. Lines, J. Appl. Phys. 41 (1970) 533.
27. S. C. Abrahams, P. Marsh, Acta Cryst. B42 (1986) 61.
28. D. M. Smyth, Proc. Intern. Symp. Appl. Ferroelectrics, 1986 (1986) 115.

Two-Wave Mixing Photorefractive Diffraction Efficiency

G. C. Gilbreath
Naval Research Laboratory
4555 Overlook Ave., SW
Washington, D.C. 20375

and

F.M. Davidson
Johns Hopkins University
Dept. of Electrical and Computer Engineering
Baltimore, MD 21218

ABSTRACT

In this paper, a new figure of merit for steady-state diffraction efficiency for two-wave mixing in photorefractive materials is presented which includes total incident energy and absorption losses. The parameter is appropriate for use in systems where the reading and writing beams are the same and where efficient use of a fixed light budget is a design requirement. Design relationships between the exponential gain, Γ, the beam ratio, m, energy coupling, absorption, and Fresnel reflectance and transmittance are explored. Experimental results using $BaTiO_3$ at $\lambda = 514.5$ nm bear out analysis. The results provide a straightforward guide to the optical designer who wishes to use the coupling capabilities of photorefractive media in an optical system or subsystem.

I. INTRODUCTION

Holographic recording in photorefractive materials, first applied in 1968[1], is the subject of continued exploration for applications in image enhancement, real-time spatial light modulation and amplification. The latter application employs the energy transfer properties of the photorefractive effect wherein energy from a strong "pump" beam is transferred to a weaker "signal" beam[2]. This property, which has come to be known as "two-wave mixing," holds promise as an effective energy coupler for systems where efficient use of a fixed light budget is an over-riding design parameter, such as in a spacecraft laser link[3]. In employing photorefractive coupling when efficiency is a requirement as well as amplification, the diffraction efficiency-gain relationship becomes important. It will be shown that conditions which enhance gain do not necessarily enhance diffraction efficiency.

In some system contexts, present efficiency parameters which instruct the designer to enhance exponential gain in the context of definitions which are borne from static holography can lead to misleading interpretations and finally, erroneous design results. This is especially true for the case in which efficient beam coupling is the prime consideration where the reading and writing beams are the same. In this paper, a new figure of merit for two-wave mixing photorefractive diffraction efficiency, $\eta^{\alpha}_{2-\lambda}$, is defined for the steady-state condition that includes the total incident energy, the relationships between exponential gain, beam coupling, beam ratios, and absorption losses. The design tradeoffs between geometries which enhance a given material's coupling performance against Fresnel reflectance and transmittance and self-pumping considerations are also discussed. Experimental results using $BaTiO_3$ are presented to verify the analysis.

II. DIFFRACTION EFFICIENCY, GAIN AND ENERGY COUPLING

Diffraction efficiency has been defined[4], as the ratio of the intensity of a diffracted beam to that of a reading beam wherein the reading beam is a single beam independent of the two mutually coherent writing beams used to form the grating in the photorefractive material. Although this procedure is useful as a measure of the fidelity of the holographic grating which does not change the grating's characteristics, the user is not really provided with a direct figure-of-merit to quantify efficient energy transfer with respect to the total energy usable in the system.

Implicit in the spirit of definition of diffraction efficiency in general is a figure which will inform a user of device or system efficiency directly. That is, a designer wishes to know usable diffracted energy compared to total incident energy so that a realistic assessment of power considerations may be made. Using the analogy of the photorefractive coupler to the power amplifier in the electronics world, the need for a definition which includes all incident power is that same as for a definition in electronics which specifies the amplifier ac signal against the combined dc and small ac signal input.

To that end, we define two-wave photorefractive diffraction efficiency, $\eta_{2-\lambda}$, to be the ratio of light diffracted into the signal beam to the incident light which consists of both the pump beam and signal beam. Using the notation of Kukhartev:

$$\eta_{2-\lambda} = \frac{I_{-1}}{I_{+10} + I_{-10}} \tag{1}$$

where I_{+10} and I_{-10} are incident pump and signal beam intensities at $y = 0$, respectively. I_{-1} and I_{+1} are the intensities in the signal and pump beams and are defined by Eqns. (2) and (3):

$$I_{-1} = \frac{I_{+10} + I_{-10}}{1 + m \, \exp(-\Gamma L_{\text{eff}})} \tag{2}$$

$$I_{+1} = \frac{I_{+10} + I_{-10}}{1 + m^{-1} \, \exp(\Gamma L_{\text{eff}})} \tag{3}$$

where $m = I_{+10}/I_{-10}$.

When Eqn. (1) is combined with Eqn. (2), the two-wave diffraction efficiency becomes:

$$\eta_{2-\lambda} = \frac{1}{1 + m \, \exp(-\Gamma L_{\text{eff}})} \tag{4}$$

The exponential gain, Γ, is determined by material parameters[4,5]. L_{eff} is the effective length and is dependent on the relationship of the electro-optic (E-O) axis to the physical dimensions of the crystal with respect to the incident beams. For a parallelpiped geometry such as is the case typically with BaTiO$_3$, L_{eff} defined as [6]:

$$L_{\text{eff}} = \begin{cases} \dfrac{a}{\cos\beta} & \beta < \beta' \\[2mm] \dfrac{b}{\sin\beta} & \beta > \beta' \end{cases} \tag{5}$$

where $\beta' = \tan^{-1}(b/a)$.

The product of the exponential gain and the effective length, ΓL_{eff}, is maximized through geometries and material properties. As shown in Figure 3 of Reference 6 for BaTiO$_3$, the Bragg angle of contstruction, θ_B, in combination with the proper angle between the grating vector, $\hat{k}_g$, and the c-axis can give rise to very high values of Γ. However, as will be shown, a maximized ΓL_{eff} does not ensure maximized energy transfer. Figure 1 illustrates two-wave mixing geometry and notation.

Typically of interest has been the optical amplification of the signal beam. The resulting gain, G, is defined as the signal intensity in the presence of the pump divided by the signal intensity with no pump present, i.e., $I_{-1(\text{pump})}/I_{-1(\text{no pump})}^7$. This relation can be expressed as:

$$G = \frac{(1 + m)}{1 + m \, \exp(-\Gamma L_{\text{eff}})} \tag{6}$$

The parameter informs the designer as to the extent of the amplification of the signal where the signal may be spatial or temporal or both. However, the parameter gives no information regarding energy coupling with respect to the total energy in the system.

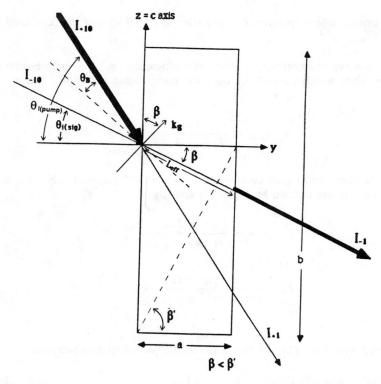

FIG. 1. Geometrical notation of "two-wave mixing" in parellelpiped nonlinear photorefractive material is presented. The wave vector k_g, is normal to the grating formed by the incident pump and signal beams, I_{+10} and I_{-10}, respectively. I_{-1} is the signal beam which acquires some of the pump energy through coupling, where I_{+1} is the pump beam after depletion. β is the angle of k_g to the c-axis; L_{eff} is the effective length; θ_B is the Bragg angle and $\theta_{i(sig)}$ and $\theta_{i(pump)}$ are the angles of incidence of the signal and pump beams.

In most cases, all of the material parameters necessary to compute analytically an optimum coupling geometry for a given material are not available except through referenced values. It is possible to find this geometry, iteratively, however, and measure values for this product directly.

ΓL_{eff} can be expressed as[4]:

$$\Gamma L_{eff} = \ln \left[\frac{I_{-1}}{I_{+1}} \frac{I_{+10}}{I_{-10}} \right] \tag{7}$$

where each of the intensities in the above expression are directly measurable as is L_{eff} from the physical dimensions and properties of the material.

A coupling coefficient, γ_c, can be defined which quantifies the transfer of energy from the pump into the signal beam as $\gamma_c = I_{-1}/I_{+1}$ and from Eqn. (7), $\gamma_c = \exp(\Gamma L_{eff})/m$. Unlike the gain, G, which increases with increasing values for m, the coupling decreases with increased m. Figures 2 and 3 exhibit these relationships for fixed values of ΓL_{eff}. Beam coupling characteristics are typically most useful for amplification applications when m is large. However, in all cases, energy coupling decreases with increasing m. Clearly, there are pump-to-signal ratios where energy coupling is small even for high values of ΓL_{eff}. Hence, conditions for large G are not necessarily conditions conducive to strong energy coupling. Present workers interested in amplification aspects of the materials choose ratios on the order of $10^4 - 10^5$. Values for ΓL_{eff} tend to range from less than 1 to 6. As can be seen from the curves, although energy is transferred from pump to signal for such parameters, i.e.: G is significant, energy transfer is small. For larger values of ΓL_{eff}, a wider range of values of m can be tolerated before coupling significantly decreases. In Figure 4, how γ_c relates to G through m is plotted for $\Gamma L_{eff} = 3.90$.

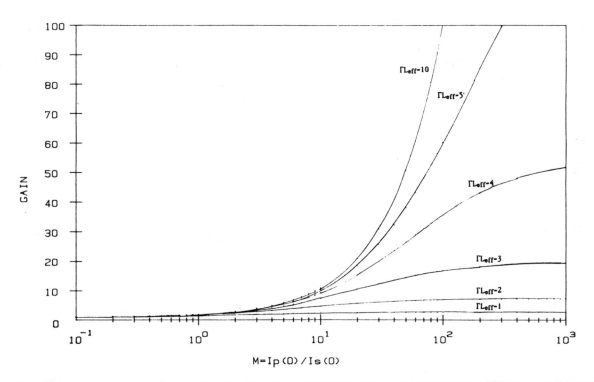

FIG. 2. Family of curves is presented for fixed values of ΓL_{eff}. The gain, $G = (m + 1)/(1 + m\exp(-\Gamma L_{eff})$, increases with increasing ratios of m, where $m = I_{+10}/I_{-10}$.

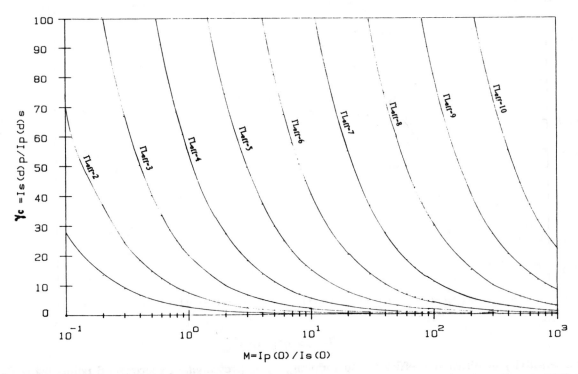

FIG. 3. Family of curves is presented for fixed values of ΓL_{eff}. The coupling, $\gamma_c = \exp(\Gamma L_{eff})/m$, decays with increasing values of m.

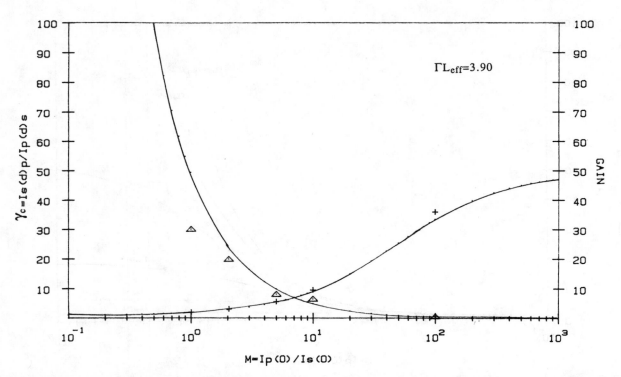

FIG. 4. Experimental points are plotted against theoretical curves for γ_c vs m and gain, G, vs m. BaTiO$_3$ was employed, oriented for $\Gamma L_{eff} = 3.90$, using $\lambda = 514.5$ nm and polarized in the plane of incidence.

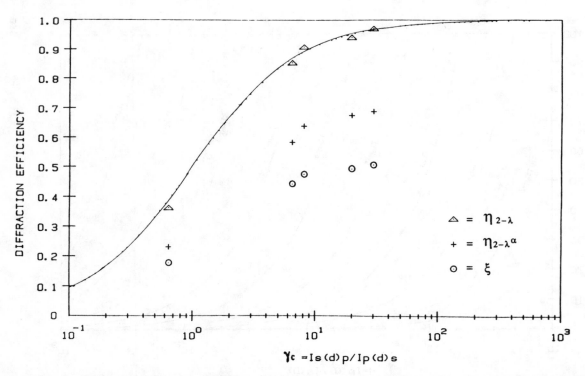

FIG. 5. The sensitivity of diffraction efficiency to coupling, γ_c, is presented. Experimental points using BaTiO$_3$ at $\Gamma L_{eff} = 3.90$ are plotted as shown for $\eta_{2-\lambda}$, $\eta_{2-\lambda}^{\alpha}$, and ξ. The latter parameter describes overall system efficiency which includes the effects of absorption, Fresnel reflection and self-pumping.

The sensitivity of $\eta_{2-\lambda}$ to γ_c can be seen in Figure 5, and is expressed as:

$$\eta_{2-\lambda} = \frac{1}{1 + (1/\gamma_c)} \tag{8}$$

As can be seen from the curves, values of $\eta_{2-\lambda}$ remain greater than 90% when γ_c is as low as 10. Referring back to Fig. 2, achieving values of γ_c on this order lower the ΓL_{eff} requirement. The gain, G, computes to 6.73 for these values, indicating that amplification is also enabled. However, for photorefractive materials where achievable values for ΓL_{eff} are generally smaller, careful choice of m is necessary to achieve efficient diffraction efficiency as well as desired amplification. Figure 6 illustrates the relationship between diffraction efficiency and gain. The experimental results plotted in this and the other figures will be discussed in Section IV.

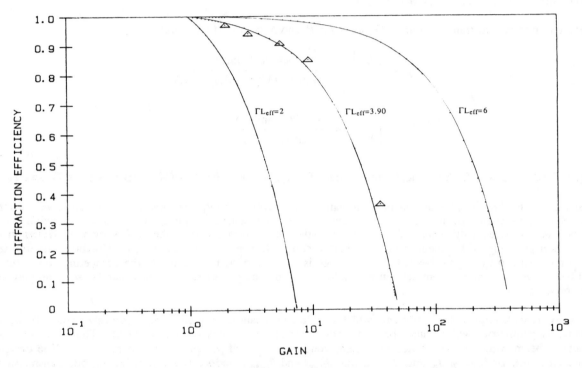

FIG. 6. Curves for diffraction efficiency, $\eta_{2-\lambda}$, are plotted against gain, G, as discussed in the text. Values for ΓL_{eff} vary as shown and indicate how these two parameters change for fixed values. Experimental data is plotted against $\Gamma L_{eff} = 3.90$.

III. FRESNEL REFLECTANCE, TRANSMITTANCE AND ABSORPTION

Reflection and transmission characteristics of polarized light incident on a given material of index of refraction, n, are well-known phenomena[8]. Although mentioned in passing as a consideration which must be taken into account in the general photorefractive literature, the relationships have not been explored from a design perspective. The effect of Fresnel reflection, in particular, is of interest. The photorefractive effect occurs in anisotropic materials where the polarization of incident light significantly influences the magnitude of the effects[9]. Consequently, from a perspective of system efficiency, it is important to consider $\theta_{i(sig)}$ and $\theta_{i(pump)}$ not only in terms of θ_B and β, but in terms of angles which reduce reflection and enhance transmission.

The equations for normalized reflection coefficients for P-polarized and S-polarized light respectively are[8]:

$$R_P = \left| \frac{E_r}{E_i} \right|^2 = \left| \frac{\tan(\theta_i - \theta_t)}{\tan(\theta_i + \theta_t)} \right|^2 \tag{9}$$

$$R_S = \left| \frac{E_r}{E_i} \right|^2 = \left| \frac{\sin(\theta_i - \theta_t)}{\sin(\theta_i + \theta_t)} \right|^2 \tag{10}$$

E_r and E_i and reflected and incident wave amplitudes. The angle of transmittance, θ_t, is computed from the incident angle, θ_i, using Snell's Law. On-axis reflection coefficients reduce to $R_P = |(n_2 - n_1)/(n_2 + n_1)|^2$ and $R_S = |-(n_1 - n_2)/(n_1 + n_2)|^2$.

Similarly, normalized transmission coefficients for P and S polarized light are:

$$T_P = \left| \frac{E_t}{E_i} \right|^2 = \frac{\sin(2\theta_i)\sin(2\theta_t)}{\sin^2(\theta_i + \theta_t)\cos^2(\theta_i - \theta_t)} \tag{11}$$

$$T_S = \left| \frac{E_t}{E_i} \right|^2 = \frac{\sin(2\theta_i)\sin(2\theta_t)}{\sin^2(\theta_i + \theta_t)} \tag{12}$$

Figure 7 plots R_P and R_S for typically used values for n_{ext} and n_{ord} for $BaTiO_3$[5]. Figure 8 plots T_P and T_S.

Considering the Fresnel phenomena in combination with geometric requirements to achieve large values for ΓL_{eff}, it appears that large angles of incidence which enhance photorefractive coupling, also reduce reflection. However transmittance is not necessarily enhanced with large angles of incidence due to the well-known phenomenon of self-pumping[10]. Self-pumping is the phenomenon whereby a conjugate beam forms due to self-diffraction. Energy transfers from the pump beam to the signal beam but this energy is then, in part, transferred to the competing conjugate. Thus, the signal beam does not enjoy a complete amplification from the pump over time. Consequently, self-pumping becomes a parasitic process.

In materials with large E-O coefficients which enable significant energy coupling, such as $BaTiO_3$, the phenomena is aggravated by geometries which cause the formation of reflection gratings and other effects. Thus, for the geometries which improve beam coupling and reduce reflection, conditions for self-pumping are also enhanced. The design trade-off then becomes one of finding θ_B and β such that $\theta_{i(\text{sig})}$ and $\theta_{i(\text{pump})}$ induce reduced reflection but transmission is not significantly affected.

To include the effects of absorption, the intensities, I_{+1} and I_{-1}, are corrected when defined as:

$$I^{\alpha}_{\pm 1} = I_{\pm 1} \exp(-\alpha L_{\text{eff}}) \tag{13}$$

where α is the absorption coefficient.

The value of α for a specific material can be found experimentally from conservation of energy considerations. In order to use the Fresnel coefficients, which make this calculation straightforward, the expression is normalized to incident power. That is

$$1 = T + R + (P_\alpha / P_i) \tag{14}$$

T is directly measurable as the ratio of the transmitted power to the incident power, P_T/P_i. For ease of computation, the on-axis geometry should be used. R is computed from the on-axis Fresnel relationship discussed in Section IV. The numerical value for α is found from $P_\alpha = \alpha L_{\text{eff}} P_i$.

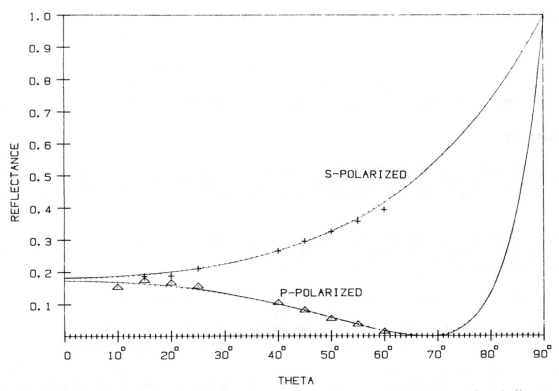

FIG. 7. Fresnel reflectance curves are plotted for $n_{ext} = 2.424$ and $n_{ord} = 2.488$. Data points indicate $n_{ext} = 2.5$ and $n_{ord} = 2.6$.

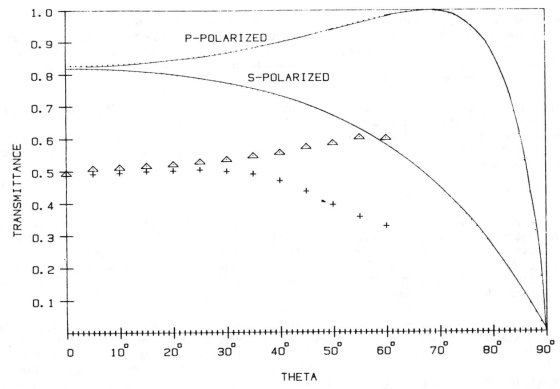

FIG. 8. Fresnel transmittance curves are plotted for $n_{ext} = 2.424$ and $n_{ord} = 2.488$. Data points plotted as Δ are transmission coefficients affected by absorption in the crystal used. Data points plotted as + indicate the effects of self-pumping in addition to absorption.

Now, the figure-of-merit for two-wave diffraction efficiency becomes:

$$\eta_{2-\lambda}^{\alpha} = \frac{1}{\exp\ (\alpha L_{\text{eff}})\ +\ m\ \exp\ [(\alpha\ -\ \Gamma)L_{\text{eff}}]} \tag{15}$$

It is clear from this expression that absorption can have a significant effect on overall system efficiencies.

IV. EXPERIMENT

Experimental data was taken using $BaTiO_3$. Data was taken using an Ar+ laser, $\lambda = 514.5$ nm, with total intensities on the order of 5-23 mW. Light was polarized parallel to the plane of incidence except where S-polarization is indicated. The c-axis of the crystal used was parallel to a rectangular face, which is typical of most $BaTiO_3$ available today.

Geometry prevented direct measurement of the Brewster's angle so points were measured and plotted as shown in Figure 7. From these values, n_{ext} was found to be 2.45, and n_{ord} to be 2.46. The Brewster angle computed to 67.8°. The absorption coefficient for this crystal was found to be .102 per mm.

Using Fainman, et al.'s, paper as a guide, the crystal was oriented for a ΓL_{eff} of 3.90. This value corresponded to a geometry of $\theta_B = 3.66°$ in the medium and $\beta = 30°$, translating to incident angles of P-polarized light of $\theta_{i\,(\text{sig})} = 39°$ and $\theta_{i\,(\text{pump})} = 21°$. The intensity ratio was varied and the experimental points were plotted in figures as shown. Direct measurements were made for all configurations of I_{-1}, I_{+1}, I_{-10}, I_{+10}, $I_{-1(\text{no pump})}$, and $I_{+1(\text{no pump})}$.

In Figure 4, experimental points are plotted for γ_c vs m and G vs m against analytic curves for $\Gamma L_{\text{eff}} = 3.90$. This figure clearly show that for values of m which enhance amplification in terms of a strong pump beam, i.e., for large G, the coupling is reduced. In Figure 5, the $\eta_{2-\lambda}$ -G relationship is plotted for three values of ΓL_{eff}. Experimental points are plotted as well.

We experimentally determined the effects of self-pumping by observing the profiles of the signal beam without the pump over time and the pump beam without the signal over time. In Figure 8, the effects of self-pumping are shown in the transmission coefficients against the angle of incidence. The first set of experimental data plots transmission affected by absorption only. The second set of data plots transmission taken over time where self-pumping clearly indicates an increased parasitic effect with increased values of θ_i.

Data points for $\eta_{2-\lambda}$ which were corrected for the measured values of α, R and self-pumping are plotted against the theoretical curve in Figure 6. In this figure, absolute efficiency, ξ, as well as $\eta_{2-\lambda}^{\alpha}$, is also plotted for this geometry. A comparison of the experimental data plotted in the figure with the analytical curve shows the values for $\eta_{2-\lambda}$ are 90% or higher for $m < 5$. However, losses due to absorption, reflection and self-pumping combine to significantly degrade overall system efficiencies.

V. CONCLUSIONS

We have shown that conditions which optimize the exponential gain, Γ, in photorefractive media, do not necessarily ensure large energy coupling. For materials which enable large ΓL_{eff} products, either through material parameters, or geometries, or both, a wider range of values of the beam ratio, m, can be tolerated and still result in good energy coupling. Since conditions which enhance gain, G, hence, amplification, will not necessarily yield efficient coupling, careful choice of m is necessary when using materials whose photorefractive properties are limited and/or for a choice of geometries which result in small values of ΓL_{eff}.

We have defined a new figure of merit to describe diffraction efficiency for two-wave mixing, $\eta_{2-\lambda}^{\alpha}$. This parameter provides a quantitative measure of system efficiency in terms of usable diffracted optical power to the total power incident on the material. This parameter is somewhat insensitive to energy coupling in that low values for γ_c can be tolerated and still yield high values for η. Hence with careful choice of m with respect to ΓL_{eff}, amplification can be significant and diffraction efficiency is not necessarily sacrificed.

Methods of experimentally determining Fresnel reflectance and transmittance and absorption are presented. Design tradeoffs are discussed in terms of geometries which enhance photorefractive performance, reduce the effects of reflection and enhance transmittance. Due to self-pumping, the Brewsters' angle is not necessarily the angle of choice even if it were achievable from the geometry of the crystal.

Significant losses in optical power are due to effects from reflection and absorption. As has been discussed in the literature, use of index-matching fluid is one way to reduce the impact of the former effect. Absorption losses are a function of the method of manufacture of the material and the choice of the material. Work is being conducted with impurities to enhance photorefractive properties. However, consequences of doping may result in changes in the absorption coefficient resulting in higher optical densities and finally, less efficient use of a fixed light budget.

VI. REFERENCES

1. F.S. Chen, J.T. LaMachia, D.B. Frazer, Appl. Phys. Lett. **13**, 223 (1968).

2. D.L. Staebler, J.J. Amodei, J. Appl. Phys. **43**, 1042 (1972).

3. G.C. Gilbreath, F.M. Davidson, Int'l Conf. on Dig. Sig. Proc., Proceedings, Sept. (1987).

4. N.V. Kukhtarev, V.B. Markov, S.G. Odulov, M.S. Soskin, and V.L. Vinetskii, Ferroelectrics, **22**, 949 (1979).

5. J. Feinberg, D. Heiman, A.R. Tanguay, Jr., R.W. Hellwarth, J. Appl. Phys., **51**, 1297 (1980).

6. Y. Fainman, E. Klanchik, S.H. Lee, Opt. Eng., **25**, 228 (1986).

7. A. Marrakchi, J.P. Huignard, Appl. Phys., **24**, 131 (1981).

8. M. Born and E. Wolf, *Prin. of Optics*, 2nd Ed., MacMillan, New York (1964).

9. J. Feinberg, J. Opt. Soc. Am., **72**, 46 (1982).

10. K.R. MacDonald, J. Feinberg, J. Opt. Soc. Am., **73**, 548 (1983).

SESSION 2
(continued)

Electro-Optics and Nonlinear Conversion II

Chair
Peter Bordui
Crystal Technology

Beta Barium borate as an electro-optic material for high power lasers*

Chris A. Ebbers

Lawrence Livermore National Laboratory, University of California
P.O. Box 5508, L-250, Livermore, California 94550

ABSTRACT

The potential of Beta-Barium Borate (β-BaB_2O_4) for use as an EO material is evaluated. The clamped and unclamped values of r_{yyy} and r_c are $|r_{yyy}^T| = 2.5$ pm/V, $|r_c^T| = 0.17$ pm/V, $|r_{yyy}^S| = 0.24 \, r_{xyz}^S$(KDP), and $|r_c^S| = 0.013 \, r_{xyz}^S$(KDP), where $r_c \equiv r_{zzz}-(n_o/n_e)^3 r_{xxz}$. The magnitude of the EO coefficient is due mainly to the electronic nonlinearity. The expected half-wave voltages and thermo-mechanical parameters of β-BaB_2O_4 plates in various configurations are calculated and compared with those of other materials of interest. Its resistance to thermal fracture, high damage threshold, and its wide range of transparency, make β-BaB_2O_4 an excellent candidate for use in a high average power Pockels cell, and as an intra-cavity laser Q-switch.

1. INTRODUCTION

As the search for designs extending the average power of lasers continues it is necessary to re-evaluate the materials used in optical elements such as frequency conversion crystals and electro-optic (EO) switches. Current lasers typically use KH_2PO_4 (KDP), deuterated KDP, or $LiNbO_3$ for these elements. BaB_2O_4 (BBO) is a new nonlinear optical material which has become available recently, and is considered an excellent crystal for second harmonic generation in the ultraviolet. In this paper we evaluate BBO for use as an electro-optic switch material for high average power lasers.

2. FUNDAMENTAL MATERIAL PARAMETERS FOR EO SWITCHES

Table 1 presents values of the parameters which are needed in order to compare the relative merits of materials used for high average power EO switches. Optical absorption is of primary importance since it determines the thermal load placed in the switch which must be removed by cooling the plate of material. With a given thermal load, the thermal conductivity will determine the temperature difference between the center of the plate and the plate surfaces used to cool the EO switch. The absolute temperature difference between the plate center and plate surfaces which a material can withstand before failing is determined by the fracture strength.

Other parameters not related to the thermal load are the EO coefficient and the damage threshold. The electro-optic coefficient determines the voltage needed in order to rotate the polarization of the incoming light by 90 degrees. Finally, the damage threshold determines the peak power a material can withstand. This parameter is relevant for lasers

Table 1. Parameters Needed to Evaluate a High Average Power Material

	Optical absorption (1 μm) α (m)	Thermal conductivity Ko (W/mK)	Fracture strength Kc $(MPam)^{1/2}$	Electro-optic coefficient r (pm/V)	Damage threshold (1 ns, 1μm) J/cm^2
Barium borate	0.01	0.8	0.15	r_{yyy} = 2.1	13.5
Deuterated KH_2PO_4 (94%)	0.4	2	0.09	r_{xyz} = 24.3	5-6
Lithium Niobate	0.1	0.1	0.14	r_{yyy} = 3.1	5-7
Quartz	0.006	5.8	1.26	r_{xxx} = 0.23	10-12

*Work performed under the auspices of the U.S. Department of Energy by Lawrence Livermore National Laboratory under Contract No. W-7405-ENG-48.

which produce high intensity pulses with a fast repetition rate. While the electro-optic coefficient and thermal conductivity are intrinsic parameters, the absorption coefficient and damage threshold more often represent the level of impurities and defects in a crystal. These latter parameters often improve with advances in crystal growth and material preparation. For example, the optical absorption at 1.064 microns in $LiNbO_3$ can sometimes be reduced by annealing the boule in an oxygen atmosphere.[1] Of course, the design of an EO switch for a particular laser must be predicated on the absorption and damage parameters of presently available materials.

3. EO FIGURES OF MERIT

In Table 2, the EO switching figure of merit, $n_o^3 r$, for each material is presented, as well as the configuration of the device. It is most advantageous to propagate down the optic axis of a crystal, if crystal symmetry permits, to avoid having to compensate for the natural birefringence of the switch. This also eliminates the problem of thermo-optic depolarization due to differences in dn_o/dt and dn_e/dt. Operating the switch in a transverse configuration, longer propagation length vs. length across which the voltage is applied, allows the switching voltage to be modified.

Table 2. Switch Configuration and Figure of Merit

	$n_o^3 r^3$ (pm/V)	$\vec{K}$	$\vec{E}$
Deuterated KDP	80.8	[110]	[001]
$LiNbO_3$	34.5	[001]	[010]
Barium borate	7.7	[001]	[010]
SiO_2	0.83	[010]	[100]

$$V\pi \propto 1/n_o^3 r$$

The clamped (relevant for fast switching frequency) electro-optic coefficients of BBO have been recently measured in Ref. 2. The unclamped (static) values were also measured. The measured value for the unclamped coefficient r_{yyy} is in good agreement with the value reported in Ref. 3. There is disagreement on the value of r_c, however there is agreement that it is an order of magnitude smaller than r_{yyy}.

4. THERMAL FRACTION AND POWER LIMITS

The thin plate fracture temperature of the four materials has been recently calculated in Ref. 4., presented in Table 3. The thin plate fracture temperature of BBO is significantly higher than deuterated KDP or lithium niobate. From this, the average power limit of a plate with a 1cm x 10cm input aperture can be calculated. The power limit is proportional to the fracture temperature, the aspect ratio and the thermal conductivity in the direction of the thermal gradient. It is inversely proportional to the optical absorption. The average power limit of BBO is a factor of 10 higher than lithium niobate.

Table 3. Thin Plate Fracture Temperatures, Average Power Thresholds, and Switching Voltage

	ΔT_F (°C)	ω_{Th} (kilowatts)	V_π (kilowatts)
Deuterated KDP (94%)	3.5	1.4	2.6
$LiNbO_3$	4.5	20.1	6.1
Barium borate	40	256	27.6
SiO_2	110	7600	256

ΔT_F ≡ thin plate fracture temperature
ω_{Th} ≡ average power threshold for 1 cm x 10 cm x 5 cm plate
V_π ≡ switching voltage required for above plate

5. SWITCHING VOLTAGE AND DEVICE SIZE

The voltage needed to use a material as an EO switch is calculated for a plate with a propagation length five times greater than the thickness in the direction in which the field is applied (Lx = 1 cm, Ly = 10 cm, Lz = 5 cm). The characteristic switching voltage of BBO is an order of magnitude less than quartz, which needs 256 kV. It should be noted that the use of quartz as an EO switch is not impractical and has been recently demonstrated in Ref. 5. A 360 kV, 10 ns pulse was applied to the (100) faces of piece of quartz with dimensions Lx = 1cm, Ly = 4 cm, Lz = 4 cm. This voltage was sufficient to rotate the polarization of a helium-neon laser by 90 degrees as it propagated in the [010] direction, in a double pass configuration. To prevent electrical flashover of the air across the electrodes, the switch was surrounded by 5 atmospheres of SF_6 gas.

The limiting power depends upon the energy deposited as heat/unit volume of material and is therefore independent of the propagation length. To reduce the voltage required to use BBO as a switch to that of lithium niobate requires a propagation length 4 times greater. Even so, the power loss due to optical absorption in BBO will be equal to or less than the power loss in a piece of lithium niobate with an equivalent half-wave voltage.

6. CONCLUCIONS

The two constraints of high peak and high average power puts a stringent limit on the minimum volume needed for EO switch materials. KDP, $LiNbO_3$, SiO_2 are available in large sizes. Large crystal growth of BBO has yet to be demonstrated. However, if it were then this material has clear advantages. We have determined that BBO has a higher limiting average power than lithium niobate or deuterated KDP. The damage threshold of BBO is higher than that of lithium niobate allowing it to withstand higher peak powers. BBO can be configured as an EO switch with a reasonably low switching voltage.

7. ACKNOWLEDGEMENTS

I would like to thank David Eimerl and Stephan Velsko for their comments and suggestions, both theoretical and experimental.

8. REFERENCES

1. Private communication with Mary Norton, LLNL.
2. C. A. Ebbers, APL, 52, (23).
3. H. Nakatani, APL, 52, (16).
4. D. Eimerl, Ferroelectrics, 72, (95).
5. L. Weaver, "Results of Quartz Pockels Cell Demonstration Experiment", internal memorandum HAP 87-140/4095W.

Flux growth of beta-barium borate

B.H.T. Chai, D.M. Gualtieri

Allied-Signal Inc., Electronic Materials and Devices Laboratory
P.O. Box 1021R, Morristown, New Jersey 07960

and M.H. Randles

Allied-Signal Inc., Snythetic Crystal Products
P.O. Box 410168, North Carolina 28241

ABSTRACT

β-BaB_2O_4 has recently been demonstrated as a promising material for both SHG and OPO applications, particularly in the UV region. We have successfully grown β-BaB_2O_4 in a novel flux composition consisting of NaCl and Na_2O. Crystal growth rate was very fast in a pure NaCl flux with well developed facets. Na_2O was added as a retardant to slow down both the growth rate and spurious nucleation. The crystal habit also changed from long-prismatic shape to more equant semispherical. As the Na_2O concentration increased, the crystal clarity was also reduced because of more severe flux inclusion. We believed that this was due to the contamination of carbonate in the flux since Na_2CO_3 was used as a precursor for Na_2O.

1. INTRODUCTION

The low temperature form of beta-barium borate (β-BaB_2O_4 or BBO for abbreviation) has recently received a lot of attention as a useful non-linear optical material [1]. Its important properties [2,3] include large birefringence (Δn = 0.118, phase matchable from 410 to 3500 nm), high optical transparency (down to 200 nm), large nonlinear coefficient (3 to 6 times that of KDP) and high optical damage threshold (> 10 GW/cm^2). It is useful in both frequency conversion (such as SHG) and parametric oscillator [4,5] applications particularly in the UV region. Indeed, fifth harmonic generation of Nd:YAG laser light has been reported [6-9].

Barium borate was first synthesized in 1874 through the reaction of NaB_2O_4 and $BaCl_2$, giving crystals of BaB_2O_4 in a solidified mass of NaCl[10]. In 1949, Levin and McMurdie [11,12] studied the BaO-B_2O_3 system and found that BaB_2O_4 melts congruently at 1095°C as a centrosymmetric crystal. They also noticed the existence of two structure modifications (the high and low forms) and a reversible phase transition between them. The structure of the high-temperature form (α-BaB_2O_4) was determined by Mighell et. al. [13] to be rhombohedral with space group $R\overline{3}c$ and a = 7.235Å, c = 39.192Å. The structure of the low-temperature form (β-BaB_2O_4) was first determined incorrectly by Hubner [14] to be monoclinic with a centrosymmetric space group of C2/c. The structure was redetermined by Lu et. al. at the Fujian Institute of Research, China [15], to be trigonal with a non-centrosymmetric space group of R3 and unit cell a = 12.532Å, c = 12.717Å. The acentric symmetry was confirmed by a SHG experiment. The strucure was subsequently confirmed by Frohlich [16]. The α-β phase transition is at about 925°C [14].

Because of the phase transition, β-BaB_2O_4 must be grown below the 925°C phase transition temperature. Consequently, flux has to be used to lower the melt temperature. Brixner and Babcock [17] produced highly acicular, transparent needles of β-BaB_2O_4 from mixtures of $BaCl_2$ and B_2O_3. Hubner synthesized the crystal using Li_2O as a flux [14]. More recently, crystals of β-BaB_2O_4 have been grown at the Fujian Institute of P.R.C. using a variety of fluxes including $BaCl_2$, $BaF2$, Li_2O, Na_2O, and $Na_2B_2O_4$ [18]. The best crystals were obtained from solutions with Na_2O, but at a very low growth rate. The work was duplicated and confirmed at Cornell University [19].

2. PHASE DIAGRAM FOR THE BaB_2O_4-NaCl SYSTEM

NaCl was chosen as a candidate solvent because of low cost, ease of handling and its ability to reduce borate melt viscosity. Since there was no information about this system, we first investigated the phase diagram.

The BaB_2O_4-NaCl pseudobinary phase diagram (Fig.1) was determined by differential thermal analysis techniques in an ambient atmosphere Pt crucible. Starting materials were 99%-purity boron oxide, 99.9%-purity barium carbonate, and reagent grade (> 99.9%-purity) sodium chloride. The appropriate mole ratio of the starting $BaCO_3$, B_2O_3 and NaCl powder components were mixed and placed in a Pt crucible of about 500 ml in volume. The crucible was placed in a furnace set at 750°C. The furnace temperature was then increased slowly over the course of about an hour to 900°C. During this period, there were chemical reactions among the powder components involving mainly the decomposition of the $BaCO_3$ with evolution of the CO_2 gas and subsequent reaction of BaO to B_2O_3 to form BaB_2O_4. After the reaction was completed, the furnace temperature was then set just above the anticipated liquidus and kept there for 12-16 hours for equilibration.

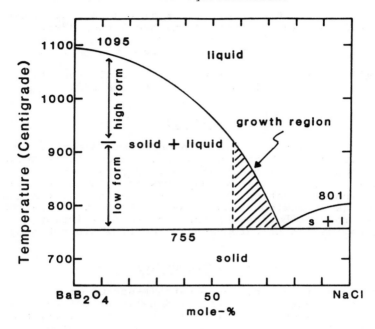

Figure 1.   Schematic phase diagram of the BaB_2O_4-NaCl system, as determined by differential thermal analysis. The highlighted area is the region for growth of beta-barium borate.

Exotherms were monitored during cooling with a differential thermocouple arrangement. The liquids of the Fig.1 phase diagram near the eutectic was investigated carefully, as were the solids and the eutectic composition. The eutectic point was determined as 25 mole% BaB_2O_4 + 75 mole% NaCl with an eutectic temperature of 755°C. Regions near the end members are interpolated. There were no phases of intermediate composition detected throughout the entire composition range. Since the β-form exists only below 925°C, this limits the melt compositions from 55 to 75 mole% NaCl for the growth of β-BaB_2O_4 as indicated in Fig.1. Since the crystallization of β-BaB_2O_4 will make the melt become increasingly NaCl rich, a starting composition with 45 mole% BaB_2O_4 and 55 mole% NaCl is preferred. It allows a maximum crystallization before the eutectic composition is reached.

3.   CRYSTAL GROWTH FROM AN NaCl FLUX

Based on the information given by the BaB_2O_4-NaCl phase diagram, a melt with 45 mole% BaB_2O_4 and 55 mole% NaCl was prepared for the growth. Needle-shaped β-BaB_2O_4 crystallized almost immediately at the melt surface when the furnace temperature cooled just below 915°C. The growth rate of these needles was extremely rapid along the long axis direction (or the c-axis [0001] direction). Crystal needles with dimensions of 1 mm x 20 mm were formed in minutes. The melt had very limited tolerance (< 10°C) of supercooling and significant supersaturation of the melt was not possible. At very slow cooling rates (< 0.3°C/hr), the needle shaped crystals can grow to larger volume with single crystals reaching both sides of the crucible (about 5 cm across).

Seeded growth was attempted using the spontaneous nucleated needle crystals as seeds. But the limited supersaturation of the melt led to extraneous nucleation on the crucible wall at the expense, or even total dissolution, of the seed. Growth on seed crystals, when it did occur, appeared to arise more from solute transport than from melt supercooling. In addition, at the growth temperature of β-BaB_2O_4, there was a slow but finite loss of the NaCl flux through evaporation. It further complicated the control of the growth rate and caused the increase of nucleation.

After several tries, we concluded that although the BaB_2O_4-NaCl system can produce β-BaB_2O_4 crystals, both the crystal shape and the extraneous nucleation make it very difficult to growth crystals of sufficient size and clarity for optical applications.

4. CRYSTAL GROWTH FROM AN NaCl-Na₂O FLUX

In order to achieve stable growth of β-BaB$_2$O$_4$ it is necessary to increase the degree of supercooling of the flux and to supress the spontaneous nucleation. We decided to add a small amount of Na$_2$O to the melt, since according to the report by Jiang [18], Na$_2$O had just the opposite effect on the growth of β-BaB$_2$O$_4$.

Seven melt compositions were prepared with the ternary combination of BaB$_2$O$_4$-NaCl-Na$_2$O. these are listed in Table I and indicated in Fig. 2. These compositions were also prepared from mixed powders and were heated and melted following the exact same procedure as with the NaCl flux. Sodium carbonate was used as the precursor for sodium oxide. Crystals of β-BaB$_2$O$_4$ were produced from all these melt compositions in the temperature range of 870 to 880°C.

Number	BaB$_2$O$_4$ mole-%	NaCl mole-%	Na$_2$O mole-%
1	52	42	6
2	54	36	10
3	46	46	8
4	48	44	8
5	47	47	6
6	46	46	8
7	66	17	17

Table 1. Solution compositions for the growth of beta-barium borate. Crystal growth occurs in the temperature range 870 - 880°C.

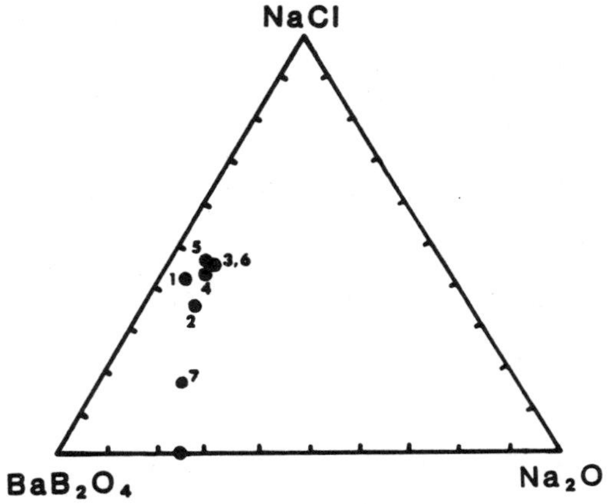

Figure 2. Compositions of solutions for the growth of beta-barium borate near 880°C. The datum in the BaB$_2$O$_4$-NaO system is from the work of Jiang, et al. (Ref. 18).

As expected, the addition of Na$_2$O to the BaB$_2$O$_4$-NaCl melt did reduce the crystal growth rate, especially along the [0001] or needle direction. The crystal morphology changed from needle or long prismatic shape to short prism or nearly spherical shape depending on the NaCl to Na$_2$O ratio. There was also an apparent reduction of the NaCl evaporation and the spurious nucleation was also greatly reduced. In addition, there was also an increase of melt surface tension which allowed better flux drainage upon the removal of crystals from the melt.

Figure 3. Photograph of beta-barium borate crystals, as spontaneously nucleated at the end of a platinum rod from a solution close to composition three in Table 1.

Although Na_2O additions seemed to have many beneficial effects, we also noticed serious drawbacks. First, there was no continuous improvement of crystal growth characteristics with further increase of the Na_2O to NaCl flux ratio. The crystal clarity was reduced with more flux inclusion. Small bubbles could be seen in the melts long after the initial fusion period, suggesting that the decomposition of sodium carbonate might not be complete and small amounts of it still retained in the flux during growth. The residual flux had a brownish color which is a further indication of retained carbonate. Fig. 3 is a photograph of the β-BaB_2O_4 crystal grown from the melt composition 3 listed in Table I.

Overall, we believe that NaCl-Na_2O is a useful flux for β-BaB_2O_4 crystal growth. A different melt preparation precedure is needed to fully react the starting powders and to remove the carbonate ion from the melt completely. A prereaction step at 1200°C similar to that suggested by Cheng et. al. [19] would be beneficial for this system.

5. CRYSTAL GROWTH FROM AN NaCl-B_2O_3 FLUX

In addition to the use of Na_2O to supress growth rate and nucleation, we also tried excess B_2O_3. The effect of excess B_2O_3 was similar to that of Na_2O, but there was a limit to about 10 mole%. Beyond this limit, a ternary composition compound became stable. A melt with 31.5 mole% BaB_2O_4, 57.3 mole% NaCl and 11.2 mole% B_2O_3 produced large clear crystals with rapid growth rate. Although a SHG test with a Q-swiched Nd:YAG laser produced intense green light comparable to that of β-BaB_2O_4, it was a totally different phase. Powder x-ray diffraction confirmed the phase to be $Ba_2B_5O_9Cl$ [20]. It had a boracite structure which was tetragonal with non-centrosymmetric space group of $P4_22_12$, and a unit cell dimension of a = 11.58Å and c = 6.69Å. Unfortunately, the crystal has no use in non-linear optics because of the extremely low birefringence ($\Delta n < 0.003$). It might be useful for electro-optic devices.

6. SUMMARY AND CONCLUSION

In search of suitable flux for the growth of β-BaB_2O_4 single crystals, we established the pseudobinary phase diagram of the BaB_2O_4-NaCl system.

Although pure NaCl flux did not appear to be suitable for β-BaB_2O_4 crystal growth due to the acicular crystal morphology and spurious nucleation, the addition of Na_2O in the flux seemed to improve the growth process.

We believe that an NaCl-Na_2O flux will be a useful flux for β-BaB_2O_4 crystal growth. However, an improved melt preparation procedure is needed to eliminate carbonate con-tamination. Careful control of the furnace temperature and the thermal gradient are also essential to the production of large, optically clear crystals.

An NaCl-B_2O_3 combination is also a suitable flux, however, care should be taken to avoid the formation of the $Ba_2B_5O_9Cl$ phase.

7. REFERENCES

1. C. Chen, B. Wu, A. Jiang and G. You, Sci. Sin., Ser B28, 235 (1985).
2. D. Eimerl, L. Davis, S. Velsko, E.K. Graham and A. Zalkin, J. Appl. Phys., v.62, 1968 (1987).
3. R.S. Adhav, S.R. Adhav and J.M. Pelaprat, Laser Focus, v.23, no.11, 88 (1987).
4 L.K. Cheng, W.R. Bosenberg and C.L. Tang, Appl. Phys. Lett., v.53, 175 (1988).
5. H. Vanherzeele and C. Chen, Appl. Opt., v.27, 2634 (1988).
6. K. Kato, IEEE J. Quantum Electron., v.QE-22, 1013 (1986).
7. K. Miyasaki, H. Sakai and T. Sato, Opt. Lett., v.11, 797 (1986).
8. W.L. Glab and J.P. Hessler, Appl. Opt., v.26, 3181 (1987).
9. C. Chen, Y.X. Fan, R.C. Eckardt and R.L. Byer, Proc. Soc. Photo-Opt. Instrum. Eng., no. 681, 12 (1986).
10. R. Benedikt, Ber. Deut. Chem. Ges., v.7, 703 (1874).
11. E.M Levin and H.F. McMurdie, J. Res. Nat. Bur. Stand., v.42, 131 (1949).
12. E.M. Levin and H.F. McMurdie, J. Am. Ceram. Soc., v.32, 99 (1949).
13. A.D. Mighell, A. Perloff and S. Block, Acta. Cryst., v.20, 819 (1966).
14. K.H. Hbner, Neues Jahrb. Mineral. Monatsh., 335 (1969).
15. S. Lu, M. Ho and J. Huang, Acta Phys. Sinica, v.31, 948 (1982).
16. R. Frohlich, Z. Krist., v.168, 109 (1984).
17. L.H. Brixner and K. Babcock, Mat. Res. Bull., v.3, 817 (1968).
18. A. Jiang, F. Chen, Q. Lin, Z. Cheng and Y. Zheng, J. Crystal Growth, v.79, 963 (1986).
19. L.K. Cheng, W. Bosenberg and C.L. Tang, J. Crystal Growth, v.89, 553 (1988).
20. T.E. Peters and J. Baglio, J. inorg. nucl. Chem., v.32, 1089 (1970).

Investigation of the Nb-rich phase boundary of LiNbO$_3$

Stephen G. Boyer
and
Dunbar P. Birnie, III

Department of Materials Science and Engineering
University of Arizona
Tucson, Arizona 85721

ABSTRACT

Changes in the point defect concentrations have been studied using Debye-Scherrer x-ray powder diffraction on samples equilibrated in various environments. The data have been integrated with information from the literature to derive an enthalpy of oxidation and an enthalpy of solution for LiNb$_3$O$_8$ into LiNbO$_3$.

INTRODUCTION

Lithium Niobate is a ferro-electric ceramic material with applications in acoustic wave devices and integrated optical circuitry. It can be grown from a melt into high quality single crystals with useful dielectric, elastic and optoelectronic properties[1]. For waveguiding applications its most notable material characteristic is the large nonlinearity in its extraordinary index of refraction. Optical waveguides can be fabricated by annealing the crystals in specific environments which form light guiding surface layers. Unfortunately, lithium niobate is susceptible to optical damage which can occur during sub-band-gap laser irradiation at room temperature. This is commonly thought to result from point defects in the crystal structure[2-4]. A better understanding of point defect mechanisms and equilibria could be instrumental in designing processing steps that inhibit optical damage.

BACKGROUND

Lithium niobate has been studied extensively. Many phase diagram, diffraction, diffusion, and optical damage studies have been performed. This section presents some work pertinent to our understanding of the niobium rich phase boundary of lithium niobate. Studies that shed light on the defect mechanisms involved with the excess niobium solution will be covered.

The phase diagram for the Li$_2$O-Nb$_2$O$_5$ system has been determined[5,6]. Figure 1 shows the impressive degree of non-stoichiometry on the niobium rich side of the phase diagram; but, there is no apparent deviation from stoichiometry on the lithium rich side. Lithium niobate melts congruently, but the congruent melting composition occurs at 48.6 mole% lithium oxide. This widely disparate behavior must be due to the point defect mechanisms that yield the non-stoichiometry.

Conductivity data is often used to elucidate point defect behavior. Figure 2 is a plot of conductivity as a function of 1/T[7]. The various lines indicate different oxygen pressure atmospheres with (a) being the most reducing and (h) the most oxidizing. The relative linearity of the plots and the good fit of the data seem to indicate that the samples were at equilibrium. Figure 3 is a plot of oxygen diffusivity as a function of 1/T by the same authors. Analyzed collectively, figures 2 and 3 seem to rule out the presence of oxygen vacancies as a majority defect since it would have taken years for the samples from figure 2 to come to equilibrium based on the diffusivities presented in figure 3; yet changes in the conductivity did occur in response to the changes in oxygen partial pressure. A second study of electrical conductivity has given similar results[8]. Figure 4 shows this other conductivity data as a function of oxygen pressure and figure 5 shows the temperature dependence. The tail end of the lines in figure 4 seem to flatten indicating a possible change in conduction mechanism. This was interpreted as a change to ionic conductivity. It is interesting to note the change in slope for some of the higher oxygen pressure lines in figure 2; these would also support the existance of a different conduction mechanism at high P_{O_2}. The slope of -1/4 in figure 4 is useful and will be derived in the discussion section. The data in figure 5 yield an enthalpy of conduction of 57.6 kcal/mole (2.50 eV).

DEFECT MODEL

Abrahams and Marsh have proposed a model that explains the non-stoichiometry in lithium niobate[9]. Detailed X-ray diffraction and density measurements of both congruent and stoichiometric single crystals were performed. By integrating the intensities of different diffraction peaks the site occupancy in the lattice was determined. The Brouwer approximation proposed involves niobium atoms sitting on lithium sites compensated by vacant niobium sites. This is consistent with the data reviewed above.

In our study of the niobium rich phase boundary, we have processed samples so that an equilibrium was reached between $LiNbO_3$ and $LiNb_3O_8$ (the adjacent phase in the phase diagram). In this case, solution of $LiNb_3O_8$ into lithium niobate is governed by the reaction:

$$3 \text{ LiNb}_3\text{O}_8 = 24 \text{ O}_o^x + 3 \text{ Li}_{Li}^x + 5 \text{ Nb}_{Li}^{\cdots\cdots} + 4 \text{ Nb}_{Nb}^x + 4 \text{ V}_{Nb}^{'''''} \tag{1}$$

This yields an equilibrium constant of

$$K_{Solution} = \left[\text{Nb}_{Li}^{\cdots\cdots} \right]^5 \left[\text{V}_{Nb}^{'''''} \right]^4 \tag{2}$$

Note that we have ignored the $LiNb_3O_8$ activity since it will equal unity when the two phases are in equilibrium. The equilibrium constant depends on the entropy and enthalpy of reaction 1 as:

$$K_{Solution} = \exp\left[\frac{\Delta S_{soln}}{R} \right] \exp\left[\frac{-\Delta H_{soln}}{RT} \right] \tag{3}$$

If we assume the above Brouwer approximation of $4[\text{Nb}_{Li}^{\cdots\cdots}] = 5[\text{V}_{Nb}^{'''''}]$, we obtain:

$$[\text{Nb}_{Li}^{\cdots\cdots}] = C \ K_{Solution}^{\frac{1}{9}}, \tag{4}$$

indicating that the slope of a plot of $\log[\text{Nb}_{Li}^{\cdots\cdots}]$ vs inverse temperature should be one ninth of the enthalpy of solution.

We have used this defect model in interpreting the present research. The following experiment was performed to determine the enthalpy of solution and further study the above defect mechanism.

EXPERIMENTAL PROCEDURE

In this study we use the fact that the variation of the lattice parameters will be indicative of the defect concentrations present. We have used lattice parameter measurements to find the solution enthaply of $LiNb_3O_8$ into $LiNbO_3$.

$LiNbO_3$ and Nb_2O_5 powders were obtained from Puratonic, both powders were 325 mesh and 6N pure. The powders were combined to obtain a mixture with twenty mole percent Nb_2O_5 and eighty mole percent $LiNbO_3$, so that after reaction the predominant phase would be $LiNbO_3$ but that significant second phase would still be present to fix the activity. Mixed powder samples were then heated in platinum crucibles for twenty-four hours with either an oxidizing or reducing atmosphere at either 900°C or 1050°C. The oxidizing environment was pure oxygen at one atmosphere, while the reducing environment was one percent oxygen in CO_2 also at one atmosphere. Following the heat treatment each sample was quenched by pulling it rapidly from the furnace.

Upon cooling, the samples were ground and put in capillaries for Debye-Scherrer X-ray diffraction. Lattice parameters were derived from the diffraction patterns using a non-linear least squares fit of both the a and c parameters using 28 reflections.

RESULTS

The lattice parameters determined in the above experiment are given in table 1. Point defects in the lattice cause small changes in these parameters. It would be difficult to quantitatively predict these changes from first principles; however, that is unnecessary since the effect on the lattice parameters is cumulative over both defects and we are merely interested in their deviation from ideality (stoichiometric composition). Changes in the lattice parameter were monitored by referencing our data to measurements of the same parameters from stoichiometric single crystals[9].

The activation enthalpy was derived by calculating the slope in $\log(a-a_o)$ versus $1/T$, where a_o is the lattice parameter for stoichiometric material. This yielded a value of 30 kcal/mole (1.3 eV). From this we can derive the solution enthalpy of $LiNb_3O_8$ into $LiNbO_3$ simply by multiplying by nine (equation 4). This yields a heat of solution of $\Delta H_{soln} = 270$ kcal/mole (11.7 eV). This is then the measured value for the enthalpy of reaction 1.

In the next section we compare this result with literature data on the shape of the Nb-rich phase boundary and discuss the oxidation and reduction behavior of $LiNbO_3$.

DISCUSSION

For comparison with the above results we refer back to the previous phase boundary determination [5]. We have replotted the Nb-rich phase boundary on an Ahrrenius plot to try to find a value for the enthalpy of solution as represented by their data. We have chosen to plot the natural log of the deviation from stoichiometry (in percent excess Nb) versus inverse temperature. This plot is given in figure 6. Just as changes in the lattice parameter are a result of a mixture of point defects that may be present, the deviation from stoichiometry has contributions from both Brouwer approximation defects. However, since both defects are constrained by overall charge balance considerations, the deviation from stoichiometry will be some scalar multiple of a relevant defect concentration. And, on a log plot, scalar quantities that multiply all points have no influence on the slope of the plot. Therefore, the slopes that we derive from figure 6 are also representative of the enthalpy of solution for equation 1.

It is interesting that figure 6 appears to have two regions. This could be representative of two different solution mechanisms, one dominating at high temperature and one dominating at low. However, we have no other explicit evidence of a transition between Brouwer regimes. We will analyze the high temperature portion of the data since at lower temperatures equilibration is slower. The high temperature data also coincide with the region over which we have measured lattice parameters. The solution enthalpy derived from the high temperature data is 260 kcal/mole (11.4 eV). This compares favorably with the enthalpy we have found using X-ray diffraction.

Referring back to equation 2 for the equilibrium constant for the solution mechanism, it can be seen that there is no factor of oxygen partial pressure. This means that, when the Brouwer approximation is defined by the defects suggested above, the niobium solubility is independent of oxygen partial pressure. We found no significant difference between samples equilibrated in different enviroments.

However, as noticed above, oxidation and reduction do occur rather rapidly in this material, notwithstanding the low values for oxygen self diffusion. To help reconcile this difference we propose the following oxidation reaction for lithium niobate:

$$\frac{3}{2} O_2 \text{ (g)} + 6e' + Nb_{Nb}^x = 3 O_o^x + Nb_{Li}^{\cdots} + 2 V_{Nb}^{'''''}$$ (5)

Oxygen is incorporated onto oxygen sites with the simultaneous transfer of niobium atoms from their normal site to the lithium site, creating niobium vacancies. This process consumes electrons which are found to be the typical conducting species ([7,8]). This reaction is chosen because the only atomic defects that play a role in this reaction are the ones that have been identified as the majority defects in the Brouwer approximation. The equilibrium constant for this reaction is:

$$K_{ox} = \frac{\left[Nb_{Li}^{\cdots}\right]^1 \left[V_{Nb}^{'''''}\right]^2}{n^6 \, P_{O_2}^{\frac{3}{2}}}$$ (6)

The concentration of the two atomic defects are controlled by the deviation from stoichiometry. In typical electrical conductivity measurements, the crystals have a constant stoichiometry and are measured as a function of temperature. For these experiments, equation 6 can be rearranged to solve for the electron concentration as a function of the oxygen pressure. The oxygen pressure slope is consistant with previous conductivity data ([8]). However a slope of -1/4 is not necessarily unique, and in fact, for this material several Brouwer approximations give this slope.

The variation in the electron concentration with temperature is then expected to be determined by minus one sixth of the enthalpy of oxidation. We can determine the variation in the electron concentration by analyzing electrical conductivity data. The conductivity σ of a material is the product of the electron concentration (n), times the electronic charge (e), times the mobility of the electron (μ):

$$\sigma = n \, e \, \mu$$ (7)

The temperature dependence of the conductivity will then be a result of the temperature dependence of both the electron concentration and the mobility. Electron conduction has been found to proceed via a small polaron mechanism with an activation energy of 0.49 eV([10]). And, the activation energy for conductivity was observed to be 2.50 eV. Then, the electron concentration changes with an activation enthalpy of 2.01 eV. Therefore, the oxidation enthalpy, ΔH_{ox}, for reaction 5 is found to be -278 kcal/mole (-12.1 eV).

This oxidation mechanism is interesting in that no macroscopic oxygen diffusion is required for the oxidation or reduction of the bulk lithium niobate. Only bulk cation diffusion is required for this process. During oxidation the oxygen atoms plate onto the surface and cations from the interior diffuse out, in the process creating the nonstoichiometry defects of Abrahams and Marsh. During reduction, oxygen is removed from the surface and cations diffuse back into the lattice causing changes in niobium site and electron concentration. The rate of oxidation and reduction can then occur at the rate of cation diffusion.

The break in the slope of the Nb-rich phase boundary seen in figure 6 may be indicative of a change in solution mechanism. The present study does not cover a low enough temperature range to verify this break in slope. If the break in slope does indeed occur then a second solution mechanism could be occuring at high temperature. Then the solution enthaply that we have found would apply to this second mechanism. One possible Brouwer approximation would be niobium antisites balanced by lithium vacancies:

$$4[Nb_{Li}^{\cdots\cdots}] = [V_{Li}']$$

(8)

This mechanism is interesting because if it did occur at high temperature, it could convert to the previous defect mechanism during normal cooling processes. This would occur by having all vacant lithium sites be filled by adjacent normal site niobium atoms leaving more niobium antisites and vacant niobiums for charge balance. This other solution mechanism is also attractive because of the rather rapid diffusion rate of lithium in the lattice in comparison with other cations (hence the ability to proton exchange at rather low temperatures ([11])).

For a given amount of nonstoichiometry, the Abrahams and Marsh reaction would seem to be favored over the entire temperature range because more defects would be created; the lower temperature findings show that the enthalpy changes favor the formation of niobium vacancies. And, since the Abrahams and Marsh reaction creates more defects, the entropy of mixing contribution would then seem to favor these defects at high temperature also. Further research is required to investigate this possibility.

If another mechanism is operative at high temperatures then a different oxidation reaction is also required. This oxidation reaction would be:

$$\frac{3}{2} O_2 \text{ (g)} + 6e' + Nb_{Li}^{\cdots\cdots} = 3 O_o^x + Nb_{Nb}^x + 2 V_{Li}'$$

(9)

It uses niobium antisites and lithium vacancies, but gives the same qualitative behavior. The oxidation enthalpy for this new reaction will be the same however, only different defects will participate in the reaction constant. Since the stoichiometry is still fixed, the conductivity change with temperature will not depend on the particular Brouwer approximation. The arguments about rapid oxidation and reduction will still apply for this other oxidation reaction.

CONCLUSIONS

We have explored the niobium rich phase boundary of lithium niobate by Debye-Scherrer X-ray diffraction. From our data we have calculated the solution enthalpy of $LiNb_3O_8$ (270 kcal/mole) and compared it to a value derived from other high temperature data (260 kcal/mole). Using our data in conjunction with that published conductivity data, we have proposed an oxidation mechanism which is consistant with the defect model proposed by Abrahams and Marsh. We have determined the oxidation enthalpy for this mechanism(-278 kcal/mole). The striking feature of this mechanism is that it does not require oxygen diffusion to accomodate rapid oxidation or reduction. Finally, we have discussed the possible existance of lithium vacancies at elevated temperatures.

ACKNOWLEDGEMENT

We are grateful for the support of NSF Grant No. ECS-8710569.

REFERENCES

1) A. Rauber; Current Topics in Materials Science, Vol 1(1978): 481-601. Edited by E. Kaldis; Amsterdam: North-Holland.

2) R.L. Holman, P.J. Cressman, and J.F. Revelli; "Chemical Control of Optical Damage in Lithium Niobate." Appl. Phys. Lett. 32(1978): 280-3.

3) E. Kratzig; "Photorefractive Effects and Photoconductivity in $LiNbO_3$:Fe." Ferroelectrics 21(1978): 635-6.

4) K.L. Sweeney, L.E. Halliburton, D.A. Bryan, R.R. Rice, R. Gerson, and H.E. Tomaschke; "Point Defects in Mg-Doped Lithium Niobate." J. Appl. Phys. 57(1985): 1036-44.

5) L.O. Svaasand, M. Ericksrud, G. Nakken, and A.P. Grande; "Solid Solution Range of $LiNbO_3$." J. Crystal Growth 22(1974): 230-2.

6) L.O. Svaasand, M. Ericksrud, A.P. Grande, and F. Mo; "Crystal Growth and Properties of $LiNb_3O_8$." J. Crystal Growth 18(1973): 179-84.

7) P.J. Jorgensen and R.W. Bartlett; "High Temperature Transport Processes in Lithium Niobate." J. Phys. Chem. Solids 30(1969): 2639-48.

8) Y. Limb, K.W. Cheng, and D.M. Smyth; "Composition and Electrical Properties in $LiNbO_3$." Ferroelectrics 38(1981): 813-6.

9) S.C. Abrahams and P. Marsh; "Defect Structure Dependence on Composition in Lithium Niobate." Acta Cryst. B42(1986): 61-8.

10) P. Nagels; "Experimental Hall Effect Data for a Small-Polaron Semiconductor." The Hall Effect and Its Applications. Plenum Press(1980): 253-80.

11) J. L. Jackel, C. E. Rice, and J. J. Veselka; "Proton Exchange for High-Index Waveguides in $LiNbO_3$", Appl. Phys. Lett. 41(1982)607-8.

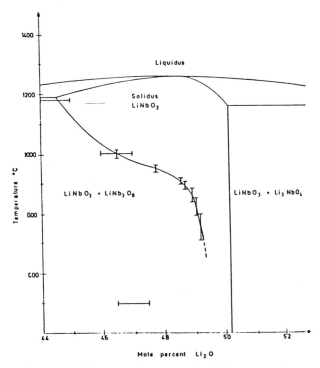

Figure 1 Phase diagram of lithium niobate showing the extent of the single phase field (from ref. 5).

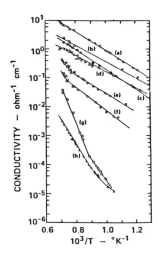

Figure 2 Electrical conductivity as a function of inverse temperature in lithium niobate for various oxygen partial pressures (from ref. 7).

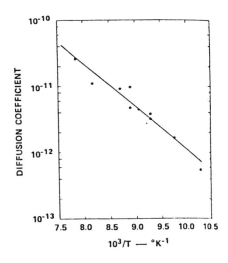

Figure 3 Oxygen diffusivity (cm^2/sec) in lithium niobate (from ref. 7).

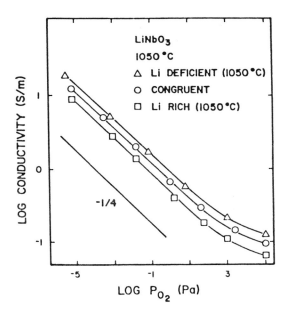

Figure 4 Electrical conductivity as a function of oxygen pressure at constant temperature (from ref. 8).

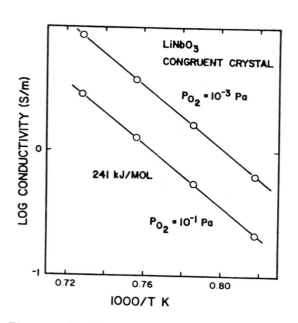

Figure 5 Electrical conductivity as a function inverse temperature at constant oxygen pressure (from ref. 8).

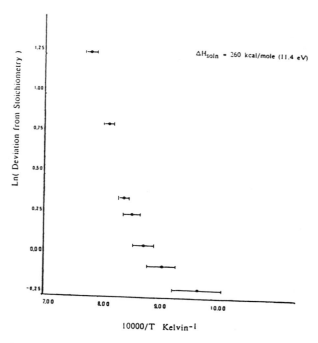

Figure 6 Temperature dependence of the deviation from stoichiometry at the Nb-rich phase boundary (data replotted from ref. 5).

Table 1 Lattice parameters determined by Debye-Scherrer X-ray diffraction.

T°C		Oxidizing	Reducing
900	a	5.1497(8)	5.1501(8)
	c	13.869(2)	13.846(2)
1050	a	5.1588(5)	5.1553(6)
	c	13.870(1)	13.828(2)

Stoichiometric Material([9]):
a = 5.14739(8)
c = 13.85614(9)

Hydrogen defects and optical damage in LiNbO$_3$

Dunbar P. Birnie, III

Department of Materials Science and Engineering
University of Arizona
Tucson, Arizona 85721

ABSTRACT

The literature pertaining to hydrogen defects dissolved in lithium niobate has been reviewed. Particular attention has been given to the infra-red absorption spectra. The polarization variations of the spectra give indications about the structure of hydrogen defects in lithium niobate. In undoped crystals hydrogen defects sit in the close-packed oxygen plane, adjacent to vacant octahedral sites that result from nonstoichiometry. In magnesium doped crystals the observed threshold effect influences the hydrogen site; at low Mg concentrations there are sufficient vacant octahedral sites, but at high concentrations the hydrogen must sit adjacent to cations and are therefore pushed out of the close-packed oxygen plane. This changes the IR spectra. The hydrogen solution model is discussed with respect to optical damage effects in lithium niobate.

INTRODUCTION

Lithium niobate is of interest for a variety of different non-linear optical applications. High quality crystals can be grown from the melt due to its congruent melting behavior. However, current applications are limited by optical damage phenomena. The primary optical damage mechanism is by the photorefractive effect (PRE). Photorefractive damage is thought to occur when the laser light ionizes trace impurities that are present in the crystals (typically iron or other transition metals). The liberated charge carriers then drift or diffuse into dark areas and are trapped. This builds a local static electric field that modifies the index of refraction through the electro-optic coefficients. This change in index of refraction causes laser beam quality reduction or dechanneling etc.

The present paper is focused on understanding techniques for lowering PRE optical damage. Specifically, several observations in the literature indicate that the presence of hydrogen lowers the optical damage level. This is important because hydrogen is frequently present in as-received crystals and can be introduced rather rapidly during different processing steps used for device fabrication.

The following sections will cover some literature that is pertinent to understanding the role of hydrogen in lowering optical damage in lithium niobate. First, the crystal structure of lithium niobate is reviewed with particular interest in how nonstoichiometry is accomodated by the crystal. Then, many reports of properties of hydrogen defects are reviewed. The infra-red absorption data are especially useful in understanding the location of hydrogen defects. Finally, the data are analyzed and discussed.

BACKGROUND

Lithium niobate is a non-centrosymmetric crystal based on a hexagonal close packing of oxygen atoms. It is symmetrically related to the Al$_2$O$_3$ structure by an ordered substitution of lithium and niobium for aluminum. This substitution is shown schematically in figure 1. The horozontal lines represent close-packed planes of oxygen ions viewed from the side. Lithium and niobium atoms replace the aluminums to give an ordered arrangement. Both lithium and niobium ions sit in quasi-octahedra of surrounding oxygen ions.

The structure and site occupancy of lithium niobate have been characterized in both stoichiometric and congruent crystals using single crystal diffraction and density measurements ([1]). The diffraction intensities in the stoichiometric material are consistent with the structure given in figure 1. In the congruent crystal the diffraction intensities and density are consistent with a nonstoichiometry mechanism of niobium antisite defects charge balanced by vacant niobium sites. The overall charge neutrality equation (the "Brouwer approximation") for lithium niobate is then:

$$4[Nb_{Li}^{\cdots\cdot}] = 5 \, [V_{Nb}^{''''''}] \tag{1}$$

where, $Nb_{Li}^{\cdots\cdot}$ represents niobium antisite defects that are charged +4 relative to the normal site ion, $V_{Nb}^{''''''}$ represents the niobium vacancies that are charged –5 with respect to the normal site atom, and the brackets indicate concentrations of these defects. Figure 2 shows how these defects would modify the structure given in figure 1. The niobium vacancies (octahedral vacancies) are adjacent to octahedral sites that are normally vacant in the structure.

The defects that are present in nonstoichiometric material are believed to be important for solution of other impurities in lithium niobate. This has been analyzed for the case of titanium solution which increases the vacant niobium concentration ([2,3]). This defect structure will form the basis for further discussion of hydrogen defects in lithium niobate.

HYDROGEN DEFECTS IN LITHIUM NIOBATE

Hydrogen is important for lithium niobate for a variety of reasons. First, it is an ubiquitous impurity; it is often present in as-grown crystals and can be incorporated during any processing heat treatments that might be performed in air. In these cases water vapor is the source of the hydrogen. Hydrogen is also important because of one technique for producing waveguides in lithium niobate: proton exchange([4,5,6]). Proton exchange is typically performed by masking areas of the surface with evaporated metal and immersing the wafer in molten benzoic acid; hydrogen atoms diffusion in while lithium atoms diffuse out, giving a surface with a different index of refraction. Hydrogen can be incorporated to rather large levels before the structure becomes unstable. At hydrogen substitutions larger than about 50% the structure becomes a cubic perovskite([7,8]); this structural transformation takes place easily, without any bond breaking required([8,9]). Hydrogen can also be introduced during the poling procedure used to give mono-domain crystals([10]). Here, the hydrogen ions drift in response to an applied electric field. All of the above processes are very rapid. Rapid diffusion rates for hydrogen in lithium niobate have been observed ([11-14]) and are certainly a result of the small size of hydrogen.

When hydrogen is introduced into lithium niobate infra-red (IR) absorption peaks can be observed([6,8,11,12,15-19]). The peaks can be interpreted as OH stretch vibrations; the hydrogen atoms sit close to the oxygen atoms in the lithium niobate lattice. An interesting observation of this IR spectrum is that the absorption parallel and perpindicular to the c-axis are different; there is no absorption when the electric vector is parallel to the c-axis([15,16,18]). Figure 3 shows IR spectra for the two orientations. This anisotropy of the absorption indicates that the OH bond is oriented in the close-packed oxygen planes. This anisotropy is also observed for OH absorption in aluminum oxide([20]). The OH peaks in aluminum oxide were also interpreted in terms of hydrogen atoms in the oxygen planes.

The initial identification of OH bonds in the close packed planes must be analyzed further. The crystal structure is such that between every two planes of close-packed oxygen layers two thirds of the octahedral sites are filled with cations; all of the tetrahedral sites remain empty. The octahedra in successive layers are stacked directly above octahedra in layers below; these adjacent octahedra each share three common oxygen atoms in a triangular arrangement. The presently proposed model of hydrogen incorporation is that the hydrogen sits somewhere in this triangle of oxygen atoms (and is therefore in the close-packed oxygen plane). This explains the polarization behavior of the IR absorption spectra for hydrogen in lithium niobate.

However, when examining the structure more closely, it will be observed that in the perfect structure (the ideally stoichiometric material), for each pair of adjacent octahedra along the c-axis only one of the sites is normally vacant; the other octahedral site holds either a lithium or a niobium atom. Therefore, in perfect material it would be expected that an hydrogen sitting among the triangle of oxygen atoms would be repelled from the cation toward the normally empty octahedral site; this would yield a structure that would still have some component of IR absorption parallel to the c-axis.

This inconsistency can be reconciled by noticing the affect of nonstoichiometry on the structure. Whenever niobium atoms are incorporated, vacant niobium sites are created. This creates situations where the two adjacent octahedral sites along the c-axis are both vacant. The present model suggests that the triangle of oxygens share by these two octahedral sites will be the preferred site for dissolved hydrogen atoms; with neither of the octahedral sites occupied by a cation the hydrogen may easily be aligned in the close-packed oxygen plane. This oxygen plane location is the third plane down from the top in figure 2.

This model for hydrogen incorporation may shed light on the observed "threshold" effect found for magnesium doping of lithium niobate[21]. Magnesium is added to lithium niobate to lower the PRE optical damage[22]. This addition of magnesium increases the photoconductivity[23], thus preventing the build-up of static fields that cause PRE[24]. It has been observed that the OH stretch peaks found in lithium niobate change as a function of magnesium concentration[19,25]. At low concentrations of magnesium doping the OH stretch peak appears in the same location as that found in pure material. When the Mg concentration reaches some threshold level, a new OH absorption peak is observed. This new absorption peak has some measureable component of absorption parallel to the c-axis, suggesting that the OH bond is now slightly inclined with respect to the oxygen planes[26]. This type of distortion could occur if the hydrogen atoms were being forced to sit in oxygen triangles that do have one of the adjacent octahedral sites occupied with some cation; the hydrogen would be repelled away from this cation giving the OH bond some c-axis component.

Then the "threshold" behavior for Mg-doped lithium niobate would correspond to the condition where the addition of magnesium had significantly depleted the supply of vacant octahedral sites. This reduction of octahedral vacancies with magnesium can be anticipated due to the lower valence of Mg cations with respect to the average cation valence of lithium niobate; this average valence is +3, while the valence of magnesium is +2. We can write a solution reaction for magnesium oxide that is compatible with the observed mechanism for nonstoichiometry in lithium niobate:

$$3\ MgO + 2\ Nb_{Li}^{\cdots\cdot} + V_{Nb}^{'''''} = 3\ O_o^x + 2\ Nb_{Nb}^x + 3\ Mg_{Li}^{\cdot} \tag{2}$$

This solution mechanism shows that as more magnesium oxide is dissolved in the lattice, niobium antisites and vacant niobium sites are consumed creating substitutional magnesium atoms and niobium atoms on niobium sites. (The choice of magnesium atoms substituted for lithium atoms is compatible with recent X-ray diffraction measurements[27].) The threshold effect found for magnesium solution in lithium niobate then must result from drastic reduction in the vacant octahedral site concentration and the resulting requirement that the dissolved hydrogen atoms must sit near some cation.

The solution mechanism for magnesium oxide suggests a similar solution mechanism for hydrogen:

$$3 \; H_2O + 5 \; Nb^{\cdots\cdots}_{Li} + 4 \; V''''_{Nb} = 3 \; O^x_o + 5 \; Nb^x_{Nb} + 6 \; H^x_{Li} \qquad\qquad (3)$$

The hydrogen defects that have been created here (represented in Kröger-Vink notation as H^x_{Li}) are not on the exact lithium location; instead, they are displaced from the center of the octahedral site into the close-packed oxygen plane. These hydrogen defects may be thought of as sitting next to a vacant octahedral lithium site (i.e. $(H_i V_{Li})^x$). Therefore, by raising the water vapor pressure, and therefore increasing the dissolved hydrogen content, the IR spectrum should have only the in-plane component. This is indeed observed for IR spectra of proton exchange lithium niobate[16]. This solution mechanism would however also cause a reduction the the vacant octahedral site concentration as more hydrogen is dissolved in the lattice.

This may help explain some observations that crystals having hydrogen are less susceptible to optical damage. Three different studies have shown reduced optical damage in the presence of hydrogen defects. First, it was found that hydrogen introduced during poling caused a reduction of the optical damage [10]. Second, optical damage in titanium in-diffused waveguides was reduced by performing a subsequent proton exchange on the wavequides [28]. And third, superior lithium out-diffused wavequides were prepared in an enviroment containing humidity; this resulted in waveguides with lower optical damage[29].

These observations of reduced optical damage may be compared with the observation of reduced damage with the addition of magnesium or in crystals that are lithium rich[19]. In all cases shown here (H, Li, and Mg additions), the addition of atoms with valence lower that 3+ has caused a reduction in optical damage. The common theme for these three additions is that all would cause a reduction in the octahedral vacancy concentration. If the average valence of all cations present is greater than 3+ (e.g. in the case of Nb^{5+} excess or Ti^{4+} doping) then octahedral vacancies must be created for overall charge neutrality in the crystal. Any of the lower valence additions will cause a lowering of the average valence in the crystal and therefore a reduction of the octahedral vacancy concentration.

Because of the similar behavior seen for Li, Mg and H additions, some photorefractive optical damage may therefore be related to the concentration of octahedral vacancies present. They may act as trapping sites for photogenerated charges in the crystal.

CONCLUSIONS

The present analysis of literature observations of hydrogen defects in lithium niobate has yielded several conclusions.

1) Hydrogen defects dissolved into lithium niobate sit adjacent to vacant octahedral sites.

2) A solution mechanism for hydrogen in lithium niobate has been proposed. This solution mechanism is compatible with the current understanding of the mechanism for nonstoichiometry.

3) This solution mechanism shows that the addition of hydrogen to the lattice will reduce the octahedral vacancy concentration.

4) The proposed solution model is compatible with the "threshold" effect observed for magnesium doped lithium niobate.

5) The comparison to other cases where optical damage is reduced suggests that <u>any</u> addition with a valence lower than 3+ should be beneficial in reducing the octahedral vacancy concentration.

6) Because the damage is lowered for a variety of different low valence additions, it is possible that vacant octahedral sites participate in the photorefractive optical damage mechanism. Conditions should be promoted that discourage formation of octahedral vacancies; the average valence of all cations present should be equal to 3+.

These conclusions result from the synthesis of a wide variety of literature reports related to optical damage and hydrogen defects. The crystal structure and mechanism of nonstoichiometry in lithium niobate has helped understand the observed polarization dependent infrared absorption in a variety of samples as a function of doping.

ACKNOWLEDGEMENT

The author is grateful to Asif A. Mufti for assistance during the initial stages of literature research. This research has been supported by NSF Grant No. ECS-8710569.

REFERENCES

1) S. C. Abrahams and P. Marsh; "Defect Structure Dependence on Composition in Lithium Niobate", Acta Cryst. B42(1986)61-68.

2) B. Guenais, M. Baudet, M. Minier, and M. Le Cun; "Phase Equilibria and Curie Temperature in the $LiNbO_3$-$xTiO_2$ System, Investigated by DTA and X-Ray Diffraction", Mat. Res. Bull 16(1981)643-53.

3) P. K. Gallagher and H. M. O'Bryan; "Effects of TiO_2 Addition to $LiNbO_3$ on the Cation Vacancy Content, Curie Temperature, and Lattice Constants", J. Amer. Ceram. Soc. 71(1988)C56-C59.

4) J. L. Jackel, C. E. Rice, and J. J. Veselka; "Proton Exchange for High-Index Waveguides in $LiNbO_3$", Appl. Phys. Lett. 41(1982)607-8.

5) D. F. Clark, A. C. G. Nutt, K. K. Wong, P. J. R. Laybourn, and R. M. De La Rue; "Characterization of Proton-Exchange Slab Optical Waveguides in z-cut $LiNbO_3$", J. Appl. Phys. 54(1983)6218-20.

6) A. Loni, R. M. De La Rue, and J. M. Winfield; "Proton-Exchanged, Lithium Niobate Planar-Optical Waveguides: Chemical and Optical Properties and Room-Temperature Hydrogen Isotopic Exchange Reactions", J. Appl. Phys. 61(1987)64-7.

7) C. E. Rice and J. L. Jackel; "$HNbO_3$ and $HTaO_3$: New Cubic Perovskites Prepared from $LiNbO_3$ and $LiTaO_3$ via Ion Exchange", J. Sol. St. Chem. 41(1982)308-14.

8) J. L. Jackel and C. E. Rice; "Topotactic $LiNbO_3$ to Cubic Perovskite Structural Transformation in $LiNbO_3$ and $LiTaO_3$", Ferroelectrics, 38(1981)801-4.

9) H. D. Megaw; "Ferroelectricity and Crystal Structure. II", Acta Cryst. 7(1954)187-94.

10) R. G. Smith, D. B. Fraser, R. T. Denton, and T. C. Rich; "Correlation of Reduction in Optically Induced Refractive-Index Inhomogeneity with OH Content in $LiTaO_3$ and $LiNbO_3$", J. Appl. Phys. 39(1968)4600-4602.

11) W. Bollmann, K. Schlothauer, and O. J. Zogal; "Bestimmung der OH^--Konzentration in $LiNbO_3$-Kristallen durch Protonenresonanz-Untersuchungen", Krist. Und Tech. 11(1976)1327-32.

12) W. Bollmann and H.-J. Stöhr; "Incorporation and Mobility of OH^- Ions in $LiNbO_3$ Crystals", Phys. Stat. Sol A39(1977)477-84.

13) R. Gonzalez, Y. Chen, K. L. Tsang, and G. P. Summers; "Diffusion of Deuterium and Hydrogen in Crystalline $LiNbO_3$", Appl. Phys. Lett. 41(1982)739-41.

14) C. E. Rice, J. L. Jackel, and W. L. Brown; "Measurement of the Deuterium Concentration Profile in a Deuterium-Exchanged $LiNbO_3$ Crystal", J. Appl. Phys. 57(1985)4437-40.

15) J. R. Herrington, B. Dischler, A. Räuber and J. Schneider; "An Optical Study of the Stretching Absorption Band Near 3 Microns from OH⁻ Defects in LiNbO₃", Sol. St. Comm. 12(1973)351-54.

16) C. Canali, A. Carnera, B. Della Mea, P. Mazzoldi, S. M. Al Shukri, A. C. G. Nutt, and R. M. De La Rue; "Structural Characterization of Proton Exchanged LiNbO₃ Optical Waveguides", J. Appl. Phys. 59(1986)2643-49.

17) L. Kovacs, V. Szalay, and R. Capelletti; "Stoichiometry Dependence of the OH⁻ Absorption Band in LiNbO₃ Crystals", Sol. St. Comm. 52 (1984) 1029-1031.

18) A. Förster, S. Kapphan, and M. Wöhlecke; "Overtone Spectroscopy of the OH and OD Stretch Modes in LiNbO₃", Phys. Stat. Sol. B143(1987)755-64.

19) D. A. Bryan, R. Gerson, and H. E. Tomaschke; "Increased Optical Damage Resistance in Lithium Niobate", Appl. Phys. Lett. 44(1984)847-9.

20) H. Engstrom, J. B. Bates, J. C. Wang, and M. M. Abraham; "Infrared Spectra of Hydrogen Isotopes in α-Al₂O₃", Phys. Rev. B21(1980)1520-26.

21) K. L. Sweeney, L. E. Halliburton, D. A. Bryan, R. R. Rice, R. Gerson, and H. E. Tomaschke; "Threshold Effect in Mg-Doped Lithium Niobate", Appl. Phys. Lett. 45(1984)805-7.

22) D. A. Bryan, R. R. Rice, R. Gerson, H. E. Tomaschke, K. L. Sweeney, and L. E. Halliburton; "Magnesium-Doped Lithium Niobate for Higher Optical Power Applications", Opt. Eng. 24(1985)138-43.

23) R. Gerson, J. F. Kirchhoff, L. E. Halliburton, and D. A. Bryan; "Photoconductivity Parameters in Lithium Niobate", J. Appl. Phys. 60(1986)3553-57.

24) E. Krätzig; "Photorefractive Effects and Photoconductivity in LiNbO₃:Fe", Ferroelectrics 21(1978)635-6.

25) L. Kovacs, I. Földvari, and K. Polgar; "Characterization of LiNbO₃ Crystals Resistant to Laser Damage", Acta Phys. Hung. 61(1987) 223-26.

26) S. E. Kapphan; "Behavior of Protons in Ternary Oxides", in Advances in Ceramics 23: Nonstoichiometric Compounds, pp379-86, 1987, American Ceramic Society, Inc.

27) H.-R. Tan and C.-F. He; "Recent Works on Optical Crystals in Shanghai Institute of Ceramics, Academia Sinica", This proceedings.

28) J. L. Jackel, D. H. Olson and A. M. Glass; "Optical Damage Resistance of Monovalent ion Diffused LiNbO₃ and LiTaO₃ waveguides", J. Appl. Phys. 52(1981)4855-6.

29) R. L. Holman and P. J. Cressman; "Optical Damage Resistance of Lithium Niobate Waveguides", Opt. Eng. 21(1982)1025-32.

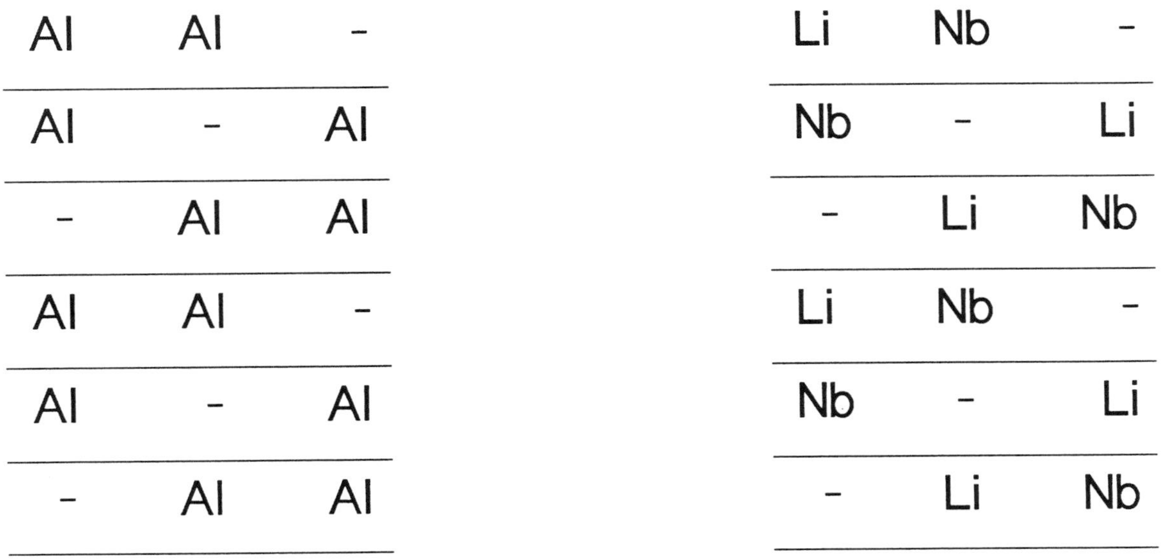

Al	Al	-
Al	-	Al
-	Al	Al
Al	Al	-
Al	-	Al
-	Al	Al

Li	Nb	-
Nb	-	Li
-	Li	Nb
Li	Nb	-
Nb	-	Li
-	Li	Nb

Figure 1 Schematic representation of the lithium niobate structure as an ordered Al_2O_3 structure. Horozontal lines represent close-packed oxygen planes viewed from the side. Aluminum atoms are replace alternately by lithiums and niobiums to give a structure with no center of symmetry. Normally-vacant octahedral sites are shown as "-" in the stacking sequence.

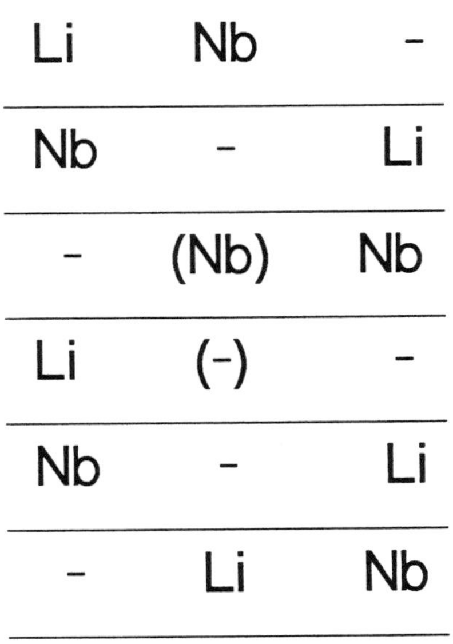

Li	Nb	-
Nb	-	Li
-	(Nb)	Nb
Li	(-)	-
Nb	-	Li
-	Li	Nb

Figure 2 Schematic representation of niobium-rich lithium niobate. Excess niobium atoms substitute for lithium and vacant niobium octahedral sites are included for charge balance.

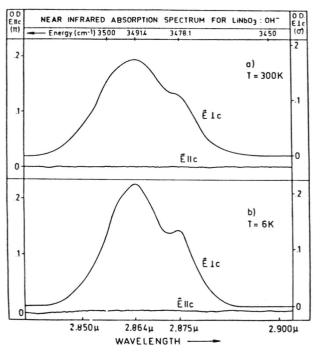

Figure 3 Observed OH absorption spectra for the two primary directions in lithium niobate. No absorption occurs for light polarized parallel to the c-axis. (From reference 15.)

Advances in the production of KNbO$_3$ crystals

G.J. MIZELL and W.R. FAY

Virgo Optics Incorporated, 6736 Commerce Avenue,
Port Richey, Florida 34668

Y. SHIMOJI

University of South Florida, Department of Physics
4202 Fowler Avenue, Tampa, Florida 33620

ABSTRACT

Techniques for producing large single domain potassium niobate (KNbO$_3$) crystals suitable for optical devices are discussed. Single crystals measuring up to 25x25x25 mm have been grown from potassium-rich melts using a top seeded technique. An x ray diffractometer and goniometer capable of orienting bulk crystals and fabricated pieces with an accuracy of 15 arc seconds was designed and constructed. An improved method of poling permits the alignment of ferroelectric domains within minutes using infrared heat. Fabrication of crystals has been streamlined through the use of special fixtures. Efficient room temperature second harmonic generation (SHG) of the 532 nm laser line is obtained using the $\langle 101 \rangle$ cut.

1. INTRODUCTION

KNbO$_3$ crystals were first grown in the laboratory almost forty years ago[1] and have been produced on a limited basis since then. This crystal is orthorhombic at room temperature and undergoes phase changes at -50, 225 and 435 degrees C.[2] Ferroelectric domains must be uniformly aligned along the [c] direction before the crystal can be used as an optical device. Potassium niobate has the highest nonlinear coefficient of any known inorganic material, and phase matches over the approximate range of 840 to 1100 nm. The production of KNbO$_3$ on a large scale has been hampered by such problems as crystal discoloration, cracking and depoling.

2. CRYSTAL GROWTH

2.1 Melt Preparation. Optically pure crystals can be grown from melts comprised of niobium pentoxide (Nb$_2$O$_5$) and potassium oxide (K$_2$O) most favorably when the potassium to niobium ratio (K/Nb) is between 1.105 and 1.273. The liquid phase is sustained at temperatures well above the melting point (MP = 1030−1045° C) for lengthy periods of time prior to crystal growth in order to avoid a blue discoloration in the crystal. The required melt soak time increases with the K/Nb ratio and varies from 5 to 25 hours. Three hundred cc platinum crucibles having a purity of 99.95% (rhodium content less than 50 ppm) are used to contain the potassium niobium oxide melts. Starting powders having a minimum purity of 99.99%, with iron and tantalum concentrations less than 10 and 50 ppm respectively, are obtained from foreign and domestic sources. Some variance in color and quality has been observed in those crystals grown from different source powders.

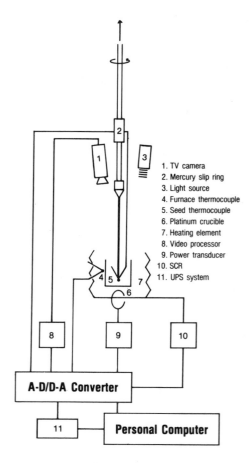

1. TV camera
2. Mercury slip ring
3. Light source
4. Furnace thermocouple
5. Seed thermocouple
6. Platinum crucible
7. Heating element
8. Video processor
9. Power transducer
10. SCR
11. UPS system

**Crystal Growth
Process Control Schematic**

Figure 1.

2.2 Furnace Design. A schematic representation of the crystal growth furnace and control system is shown in Figure 1. The furnace is designed for efficient operation at 1100 degrees C in air. Cylindrical refractory alloy heating elements are arranged and insulated to provide a favorable thermal environment for $KNbO_3$ crystal growth and cooling. Power, temperature and crystal diameter may be controlled automatically using closed loop circuits. Feedback signals can be blended and appropriately scaled using software to provide optimum control configurations and dynamic ranges.

The thermal gradient within the furnace can be modified through the use of lids. Shown in Figure 2 are the axial thermal profiles for a typical furnace measured at the growth temperature and at the two $KNbO_3$ phase change temperatures of 435° and 225° C. A notable feature of this graph is the improvement in thermal stability and axial uniformity as the furnace temperature decreases. This is beneficial toward minimizing cracking during the more destructive tetragonal to orthorhombic phase change occurring at 225 degrees C.

2.3 Pulling Mechanism. The apparatus providing pull and rotation motion for crystal growth uses a precision drive mechanism that is concentric with the pulling shaft. Pull speeds may be varied from 0.1 to 100 mm per hour, with rotation rates of 0 to 100 rpm. The mechanisms were specifically designed for smooth, vibration-free operation.

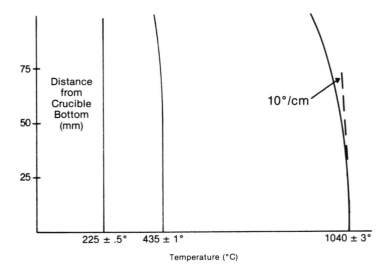

Axial Thermal Gradients in a KNbO₃
Growth Furnace

Figure 2.

2.4 Growth Technique. Growth from seeds is complicated by the two phase changes that occur during the heat-up. Parallelpipeds cut from previously grown crystals are carefully affixed to a ceramic rod which is attached to the pulling shaft. The furnace temperature is increased automatically to approximately 1040 degrees C over a period of 72 hours, in accordance with a predetermined temperature/time computer program. Seeds oriented along the orthorhombic ⟨101⟩ are used most frequently, although other orientations may be used with good results. The seed is dipped and carefully melted back to expose a surface that is free of mechanically induced defects. Seed diameter is monitored by a video position analyzer which has a detection limit of about 0.01 inch. The crystal is rotated at 30 rpm and pulled slowly, resulting in a linear growth rate of approximately 1 mm per hour. The diameter is gradually increased to between 15 and 25 mm. Once the crystal reaches the desired size it is manually separated from the melt and cooled automatically. Crystals invariably grow with pronounced { 101 } and (010) facets.

There is also great interest in the production of single crystal $KNbO_3$ fibers using the laser heated pedestal method. In one experiment, a slender seed cut from a $KNbO_3$ crystal of extremely high purity was used as feed material. Using established techniques, a 6 cm long fiber was pulled having a colorless region of about 1 cm, where the blue discoloration was avoided through control over the fiber pull speed. The growth of colorless, high quality $KNbO_3$ fibers is not yet clearly understood and is currently under study.

2.5 Cooling Technique. Cooling rates of 5 degrees per hour through the 435° phase change and 3 degrees per hour through the 225° phase change have been successfully used. Above and below these temperatures, a cooling rate of 20 degrees per hour can be used. As the furnace temperature is decreased from the growth temperature down to room temperature the crystal undergoes obvious color changes. Above 435° the crystal is deep red, turning orange as it transforms from cubic to tetragonal. Near low phase change temperature (225° C) the crystal is yellow, becoming completely colorless below 150°. We found no clear advantage in using cooling rates less than 3° per hour with our furnace system. Work is currently underway to investigate the feasibility of annealing crystals in a thermistor controlled heat pipe operating at 225° C where thermal stability can be improved by at least a factor of ten. Under these conditions cooling rates as slow as 0.1°/hr may prove effective.

3. ORIENTATION AND FABRICATION

The naturally occurring {101} and (010) facets are quickly identified by observing extinction symmetry as the crystal is rotated between crossed polarizers. Narrow flats are ground at two adjacent corners, formed by the intersection of {101} facets, corresponding to the [a] and [c] directions. These flats can be discerned quickly by x ray diffraction using a system designed and constructed in-house. This system, comprised of a Kevex miniature x ray tube (Cu Ka), a precision rotary table, a Canberra x ray detection system and a goniometer that can be transferred directly to grinding machines (Figure . 3), is inherently accurate and efficient. Some useful Bragg x ray angles are listed in Figure 4. Once the [c] direction has been identified, parallel (c) flats are ground, in preparation for the poling process. Care is taken to avoid mechanical and thermal stresses when contacting the crystal to fixtures. Precision metal fixtures used for grinding and polishing can maintain orientation to within eight minutes. The (b) faces are always polished for reasons discussed below.

Crystallographic Direction		Bragg Angle
<a>	100	15° 46'
<b>	010	22° 53'
<c>	001	15° 41'
	101	22° 36'
	111	15° 50'

Useful Bragg Angles for $KNbO_3$

Figure 4.

Figure 3. Virgo X Ray Diffractometer.

4. POLING OF FERROELECTRIC DOMAINS

4.1 Domain Formation. The unusual ferroelectric dipole mobility in $KNbO_3$ has contributed to the slow progress in developing this crystal for practical applications. Multi-domain crystals are typically transparent when viewed perpendicular to the [b] direction and flawed or cloudy when viewed along the [b]. Misaligned domains often appear as structures resulting from slip and twinning[3] or as an opalescent haze. Up to 10% of the as-grown crystals are free of visible domains.

Dipoles can be misaligned by excessive thermal or mechanical stresses. $KNbO_3$ is least sensitive to thermal depoling within the temperature range of 0 to 70 degrees C. Above 70°, care must be taken to provide uniform heating of the crystal, or a voltage of 1 KV DC per cm may be applied to the [c] faces in order to avoid partial or catastrophic depoling. Catastrophic thermal depoling generally occurs during the cool down portion of the heating/cooling cycle. Crystals may be thermally cycled from room temperature to 180° C within 10 minutes when uniform thermal coupling to the crystal is provided. No catastrophic thermal depoling has been observed when an applied voltage of 1 KV DC per cm (of [c] axis length) was maintained throughout the thermal cycle under conductive, convective and radiative thermal loading. Mechanical stresses should be limited to less than 3kg per cm². In some cases, a crystal that has been partially depoled will self heal within 24 hours in the absence of those stresses responsible for depoling.

4.2 The Poling Process. A poling technique has been described by T. Fukuda[4]. When using a similar technique in our laboratory we found poling to be unreliable. The established technique involves immersing the entire crystal and electrode assembly into silicone oil, in order to prevent arcing and to provide a uniform thermal environment. The oil bath is heated up to about 200 degrees C and a voltage ranging from 1 to 10 KV per cm is applied. While this technique is effective, the use of silicone oil is messy and can interfere with electrical flow between the electrode and crystal surface. During our experimentation we observed a strong photoconductive effect in $KNbO_3$, where an increase in current flow through the crystal was measured at constant voltage as the crystal was illuminated with a broad band light source. A method has been perfected using an intense, focussed beam of broad band light to heat the crystal. The DC voltage and length of time required to completely pole the crystal is greatly reduced using this method and the need for silicone oil has been eliminated. Crystals may be poled in bulk form or at any stage of processing, with no damage to polished surfaces or optical coatings. Advantageously, the poling process is completed within about 20 minutes.

5. SECOND HARMONIC GENERATION

The phase matching angles for the fundamental wavelength of 1064 nm have been calculated using Sellmeier equations.[5,6] We have measured efficient type 1 phase matched SHG using a ⟨101⟩ direction that is approximately 45 degrees between the [a] and [c] directions. For this case the fundamental wave is polarized in the [b] direction. A simple test set-up consisting of an iris to control beam diameter, an adjustable platform to hold the test crystal, a prism to spatially separate the 1064 and 532 nm components and an energy detector, was assembled. The Nd:YAG laser system used can deliver a 20 ns multimode pulse, at a pulse frequency of 5 hz, with a maximum energy of about 150 mj. The ⟨101⟩ $KNbO_3$ test crystal had an aperture size of 7.5 mm and a ⟨101⟩ length of 6.7 mm. This crystal was uncoated and had only a window shine on the {101} faces. When calculating conversion efficiencies, reflection losses were ignored by comparing the refracted infrared and visible energy components (V X 100% / IR + V) at the output side of the prism. For the first test a beam diameter of 6.2 mm (@ e^{-2} intensity) was used. With a pump energy of 20mj, a SHG efficiency of 10% was measured. A maximum SHG conversion efficiency of 33% was obtained with a pump energy of 52 mj. No laser damage occurred at the maximum output energy of the laser. A conversion efficiency of 47% was measured for the same crystal under similar test conditions using a 3.5 mm beam diameter, and a pulse energy of 26 mj.[7]

6. CONCLUSIONS

Large high quality crystals of $KNbO_3$ (Figures 5a and 5b) can be grown, oriented and poled on a repeatable basis. Potassium niobate possesses fairly rugged material properties when handled within certain thermal and mechanical limits. Efficient, type 1 critical phase matching by angle tuning has been demonstrated using a ⟨101⟩ cut with the fundamental wave polarized along the [b] direction. Further characterization of this material for specific nonlinear applications is necessary to determine the real potential for $KNbO_3$ devices.

Figure 5a.

Figures 5a and 5b. Typical $KNbO_3$ Crystals.

7. ACKNOWLEDGEMENTS

The authors wish to thank Mr. Jeff Dixon of Amoco Laser Corporation for his valuable technical input on $KNbO_3$ device performance, the faculty at the USF Physics Department for their continuing cooperation, Mr. Robert Thomas for his undaunted support of crystal growth and Mrs. Kim Sanchez for typing the script.

8. REFERENCES

1. B.T. Matthias and J.P. Remeika, Phys. Rev. 82 pp. 727-729 (1951).

2. E.A. Wood, "Polymorphism in potassium niobate and sodium niobate" Acta Crystallographica, Vol. 4, pp. 353-362, July, (1951).

3. W.D. Kingery, H.K. Bowen, D.R. Uhlman, Introduction to Ceramics, Second Edition, pp. 711-713, John Wiley & Sons, New York (1976).

4. T. Fukuda and Y. Uematsu, "Preparation of $KNbO_3$ Single Crystals for Optical Applications" J.J. of Applied Physics, Vol 11, No. 2, Feb. (1972).

5. J.C. Baumert, J. Hoffnagle and P. Gunter, "Nonlinear optical effects in $KNbO_3$ crystals at Al_xGa_{1-x} As, dye, ruby, and Nd:YAG laser wavelengths", spie Vol. 492, ECOOSA, pp. 374-385 (1984).

6. J. Harrison, Schwartz Electro-Optics, Inc. Private Communication.

7. M. Thomas, Sanders Associates, Private Communication.

KTiOPO$_4$ (KTP) -- past, present, and future

John C. Jacco

Ferroxcube Div. of Amperex Electronic Co. (N.A.P.C.)
5083 Kings Highway, Saugerties, NY 12477

ABSTRACT

KTiOPO$_4$ (KTP) satisfies many of the criteria required of a second harmonic generating (SHG) material. KTP combines large nonlinear optical coefficients with broad spectral, temperature, and angular bandwidths. It is also chemically inert and easy to fabricate. Accordingly it has been called the material of choice for the doubling of 1.064 μm Nd:YAG radiation. A history of the development of KTP to its present status is presented with special emphasis on crystal growth and SHG performance. A summary of various physical, chemical, crystallographic, dielectric, ferroelectric, and optical properties of the material are given. Applications such as SHG of KTP under conditions of noncritical phase matching, optical parametric oscillation, blue light generation via wave mixing, and guided wave optics are discussed. The future of KTP as a nonlinear optical material is addressed stressing solutions to crystal size limitations and bulk damage problems.

1. INTRODUCTION

The use of electro-optic devices which employ the non-zero nature of the second-order polarizability of certain crystalline materials are well known. Wave mixing schemes such as second harmonic generation (SHG), sum frequency generation (SFG), difference frequency generation (DFG), and optical parametric oscillation (OPO) had all been established shortly after the advent of the first lasers in the early 1960's[1-7]. Since then a considerable amount of effort has been devoted towards finding and developing suitable materials having the necessary properties to efficiently perform as wave mixing media in the aforementioned scenarios. The quest for high-power visible lasers has been served by utilizing the second harmonic and sum frequency technologies. Tunable sources of radiation have been produced as a result of advances in OPO development. At present, the majority of nonlinear materials used in devices have been fabricated from bulk single crystals. The stringent requirements for optical quality, transparency, efficiency, phase-matching, stability at high fluences, mechanical integrity, and chemical stability, which ideally should all coexist, limit the amount of useful materials to a minuscule pecentage of the known materials.

However, the material KTiOPO$_4$ (KTP) has been found to excell in most of the aforementioned requirements and is emerging as the leading bulk crystal for the doubling of 1.064 μm Nd:YAG radiation. Also, as crystals are becoming more readily available, KTP is being used successfully in the SFG of red, yellow, and blue visible radiation. The advantages that are offered by KTP are its large non-linear coefficients (comparable to that of Ba$_2$NaNb$_5$O$_{15}$), broad angular, spectral, and temperature bandwidths, the ability to be phase matched using either a Type I or Type II interaction, high mechanical and chemical stabilities, and a high damage threshold.

KTP was originally synthesized in 1890 by Ouvrard using molten potassium pyro- and ortho- phosphate fluxes [8]. The flux synthesis, X-ray powder data, and the Rb and Tl analogues of KTP were reported by Masse and Grenier [9]. Patents by Bierlein and Gier, and Gier described the use of KTP as nonlinear optical device, crystal growth of KTP by the flux method, and crystal growth hydrothermally, respectfully[10-12]. Zumsteg, et al.[13], then reported on the transmission and pertinent nonlinear optical parameters of the material. Since then, the body of information regarding KTP crystal growth and properties has been growing steadily. The purpose of this article is to unify this body of information into a useful reference in order to provide a picture of where KTP *has been*, *is presently*, and *will be* in the field of nonlinear optics.

2. CRYSTALLOGRAPHIC PROPERTIES AND CRYSTAL GROWTH

2.1. Crystallographic properties

The crystal structure of KTP belongs to the orthorhombic point group mm2. The habit of KTP crystals, however, appears to correspond to the point group mmm. Tordjman, et al.[14], reported the space group as being Pna2$_1$. Their space group and lattice parameter determinations were performed on flux grown KTP. Since then, lattice parameters have been reported on KTP grown by various fluxes and other methods of crystal growth. Table 1 lists some of this work.

Table 1. KTP lattice parameters.

Method of Growth	a (Å)	b (Å)	c (Å)	Ref.
Flux (Phosphate)	12.814	6.404	10.616	8
Flux (Phosphate)	12.809	6.420	10.604	15
Flux (Phosphate)	12.8164	6.4033	10.5897	16
Flux (Tungstate)	12.840	6.396	10.584	17
Hydrothermal	12.800	6.400	10.580	13

X-ray analysis also indicates that there are 8 molecules per unit cell. Powder diffraction data presented by Jacco, et al.[18], and by Cai, et al.[15], show diffraction lines not present in JCDPS file card for KTP. Anomalies in the relative intensities of certain diffraction lines were also noted. These differences have been attributed to differing methods used for crystal growth and the incorporation of impurities in the crystal. The crystal structure of KTP can be visualized as chains of alternating PO_4 tetrahedra and distorted TiO_6 octahedra. The potassium ions are situated in channels between the chains. There is also an alternating string of TiO_6 octahedra which share an oxygen atom not belonging to any phosphate ion. This Ti-O bond is shorter than the rest and is thought to be the factor responsible for the large nonlinearity of this compound.

2.2. Crystal growth

The two primary methods used for the crystal growth of KTP are the high temperature solution growth (flux) and hydrothermal methods, respectively. Since KTP melts incongruently melt growth techniques are inappropriate. To date, the largest SHG oriented pieces of KTP have been fabricated from flux grown material. Table 2 lists reported crystal growth attempts with the appropriate references.

Table 2. Crystal growth of KTP

Method of Growth	T Range (°C)	Size Crystals (mm)	Ref.
Flux	1200-900	---	9
Hydrothermal;	850-600	---	13
Hydrothermal	920-750	6x5x4	10
Flux	1000-865	15x8x2	11
Hydrothermal	600	---	19
Flux	1000-800	15x15x5	18
Flux	---	25x15x10	20
Flux	---	42.5x42x13.6	21
Flux	900-650	---	15
Flux	1000-850	8x8x6	22
Flux	1000-700	10x10x10	17
Hydrothermal	425-375	---	23
Flux	970-900	17x33x40	24

The hydrothermal method presently used for commercial production of KTP by Airtron Division of Litton Industries operates at 600 °C and 25 kpsi pressure. Gold lined autoclaves of a special alloy are employed. A six week growth run is necessary to produce a crystal large enough to yield 5 mm SHG cubes[19]. Another hydrothermal technique has been developed by Laudise, et al.[23], which operates at lower temperatures and pressures (400 °C and 10 kpsi) thus allowing the use of cheaper low carbon steel autoclaves.

The greatest progress in the growth of large KTP crystals has come from research and development in the area of flux growth. The most significant breakthrough in this endeavor has been the work of Bordui, et al.[24]. Using a patented process, large clear crystals up to 17x33x40 mm in size were grown in 10 days. Large crystals have also been produced of reasonable quality by researchers in The Peoples Republic of China using as process that takes about 40 days[20]. More recently, a system developed by Loiacono, et al.[25], at Philips Laboratories/Briarcliff, based on the work of Bordui, can routinely grow crystals 20x40x50 mm in size in 14 days from which 14 mm SHG cubes have been fabricated.

The most common solvents used in the flux growth of KTP are compositions contained in the binary phase system $(KPO_3)_n$ - $K_4P_2O_7$. The correlation between crystalline habit and both solvent composition and supersaturation has been noted by Pavlova, et al.[26]. Also, a 90° twinning phenomenon observed in crystal growth at high potassium concentrations has been reported[27]. KTP crystals have also been grown from other high temperature solvents, most notably, Ballman's tungstate solvent[17].

3. PHYSICAL PROPERTIES

The physical properties of KTP are summarized in Table 3.

Table 3. Selected KTP physical properties

Hardness	702 Knoop [28]	537 Vickers [21]	---
Density (g/cc)	2.945 [29]	2.99 [22]	3.023 [35]
Thermal Conductivity (W/cm °K)	0.13 [30]	---	---
Electrical Conductivity (c axis)	3.5×10^{-8} S/cm [31] (295°K - 1kHz)	3×10^{-5} S/cm [32] (293°K - 1MHz)	1.1×10^{-5} S/cm [33] (293°K - 1MHz)
Melting Point (°C) (Incongruent)	1167 [9]	1150 [30]	1172 [18]
Specific Heat (cal/g-°C)	0.1737 [34]	0.1643 (25°C) [35]	---

In addition to the data listed in Table 3, the dielectric properties of KTP have been measured by a number of researchers[31-33,36,37]. A dielectric anomaly has been noted at 280°K by Kalesinskas, et al.[31]. However, in generating his specific heat data, Loiacono reported seeing no thermal anomaly at that temperature[35]. Yanovskii and Voronkova have observed a dielectric relaxation phenomenon in the [001] direction between 200° and 800°C that has been attributed to high K^+ mobility in the channels of the crystal structure along the c-axis[32].

4. Optical Properties

KTP is transparent from 4.5 to .35 μm. Infrared and Raman spectral data have been reported[38-44]. An FTIR study conducted in the 2.5 to 4.5 μm range has shown differences in OH⁻ incorporation in hydrothermally vs. flux grown material[45]. Positive identification of OH⁻ absorption peaks has been made by Ahmed, et al.[46], using deuterium isotope shift techniques. The indices of refraction of KTP were originally measured by Zumsteg, et al.[13]. Since then, others have reported index of refraction measurements and several Sellmeier equations exist for calculation the index as a function of wavelength[30,47]. The accuracy of the Sellmeier fit to the refractive indices is paramount because of the need to predict phase match angles for nonlinear interactions at wavelengths other than 1.064 μm.

The nonlinear optical properties of KTP are of great interest and have been studied in some depth. Zumsteg, et al., were the first to measure d_{ij}[13]. The calculated d_{eff} using Zumsteg's data is 7.36×10^{-12} m/V. This is about 30 times that of KDP and 3 times that of LiNbO$_3$ for the doubling of 1.064 μm. Various nonlinear optical properties are presented in Table 4.

Table 4. Nonlinear Optical Properties of KTP (1.064 μm, Type II)

Spectral Bandwidth (Å-cm)	5.6 [48]	4.5-13.5 [39]	---	---
Angular Bandwidth (mrad-cm)	8.2 [50]	61 [51]	13 [49]	15 [48]
Temperature Bandwidth (°C-cm)	25 [51]	25 [48]	20 [49]	24 [47]
Walk-off Angle	0.057° [13]	0.262° [52]	---	---
Coversion Efficiency	52% [51]	68.7% [20]	65.1% [21]	59% [47]
Damage Threshold	650 MW/cm^2 [50]	>700 MW/cm^2 [20]	15GW/cm^2 [47]	1GW/cm^2 [48]

The electro-optic coefficients r_{23} and r_{33} were measured on an SHG oriented crystal by Massey, et al.[42]. The coefficients r_{13}, r_{23}, and r_{33} were also determined by Bierlein and Arweiler using X,Y, and Z oriented KTP[37]. The resulting data indicates that KTP is an attractive material for both amplitude and phase modulated electro-optic devices.

Optical degradation of KTP used in SHG and SFG experiments has been observed[50,53]. Lemeshko, et al.[54], have noticed a similar effect with the passage of an electric current along the c-axis in specimens used to measure the electro-optic effect in KTP. The current view of this phenomenon is a reduction reaction where Ti^{+4} changes valence to Ti^{+3}. Another type of occurrence which has been observed is an optical non-uniformity that is present in some flux grown material. This non-uniform area has been detected by a nonlinear visualization technique developed by Stolzenberger[49] on SHG oriented material. A more recent technique has been formulated where the native crystal boule can be scanned prior to fabrication[53]. These provide a means to "mine" homogeneous material while research continues on the identification and elimination of the non-uniformity.

5. FERROELECTRIC PROPERTIES

The possibility of ferroelectricity in KTP was alluded to by Zumsteg, et al.[13], because of the point symmetry that KTP crystallizes to. Their attempts at polarization reversal failed because of the high conductivity possessed by KTP along the polar axis. Later, Leonov, et al.[55], observed that the second harmonic signal in powdered KTP was zero at temperatures above 897°C. The signal was restored upon cooling indicating a transition. Yanovskii, et al.[36], determined the transition temperature to be 936°C by measuring the relative permittivity as a function of temperature.

Bordui has since measured the transition temperature to be 945°C[56]. Observation of ferroelectric domain structure in KTP was reported by Voronkova, et al.[57]. Domains of 180° were found by etching and SEM analysis. Bierlein and Ahmed[58] also reported domain structure in hydrothermally grown material and developed a poling process for obtaining single domain KTP. Complex domain walls were found in flux grown KTP by Loiacono and Stolzenberger[59] and are relatable to the optical non-uniformity mentioned in Section 4. The domain structure was detected by a nonlinear visualization technique, a modified pyroelectric measurement, and an optical polishing process which high-lights the surface due to differences in hardness. This domain structure is not related to the ferroelectric domains mentioned in this Section but rather thought to be similar to the Dauphine twin in quartz.

6. DEVICE APPLICATIONS

As mentioned in the Introduction, the primary use of KTP single crystals has been the frequency doubling of 1.064 μm Nd:YAG radiation. Liu, et al.[29], demonstrated 5.6 W output at 532 nm by intracavity SHG using an acousto-optically Q-switched Nd:YAG laser at 5 kHz with a Type II KTP doubler. The tripling of 1.645 μm Co:MgF$_2$ radiation to 548 nm was performed by Menyuk by summing 1.645 μm with it's SHG output[60]. Intracavity doubling of a quasi-cw Nd:YAG laser has shown that output powers greater than 25 W were obtainable for a 3x3x5 mm KTP crystal[61]. An average of 8.7 W of green light output was reported by Liu, et al.[21], at Shandong University. Driscoll, et al.[50], demonstrated efficient SHG of multimode, divergent beams. Also noted in this work was reconversion in long (9 mm) crystals which caused saturation of the conversion efficiency. This effect was eliminated by a double pass geometry with shorter (5 mm) crystals. An extracavity SHG set-up was reported by Dovchenko, et al.[62], in which highly stable, short duration picosecond pulses were produced. They were able to generate 1.064 μm with an average power of 1.9 W, a pulse power of 1.4 MW, with a pulse duration of 47 ps and also .532 μm with an average power of 1.1 W, a pulse power of 1.1 MW, with a pulse duration of 33 ps. The type of radiation produced has applications for third harmonic generation, phase-matched pumping of dye lasers, and parametric light generation.

Garmash, et al.[63], reported efficient SHG of 1.0796 μm Nd:YA (YALO) light in a Type II, noncritical phase-matched configuration. The phase-matching direction at room temperature was shown to be phi = 0°, theta = 85°20'. At a temperature of 153°C theta increases to 90°, thus achieving noncritical phase-matching. An output of 15 W of .54 μm light was obtained with a pump power of 5.5 kW and a pulse repetition rate of 5 kHz. Blue (.459 μm), coherent, cw radiation was demonstrated by Baumert, et al.[64], using noncritically phase-matched KTP by summing 1.064 and .809 μm beams. The .809 μm radiation was generated by a dye laser. Generated outputs of 0.96 mW were produced by placing KTP inside the cavity of a Nd:YAG laser that was pumped with th .809 μm output of a dye laser which also served as the source of the other fundamental. Wide angular and temperature bandwidths were also achieved. Diode laser output can be substituted for the dye laser. Yellow light at .589 μm has been generated by mixing 1.32 and 1.064 μm beams in KTP[53]. Red light at .671 μm has also been produced by SHG of the 1.3413 μm line of a Nd:YA laser[53]. The doubling efficiency was 42.6% for a 9mm long crystal. This compared to about 30% SHG efficiency for a 7 mm long BBO crystal.

KTP has been used for other types of applications than those mentioned above. The use of KTP as an OPO material has been reported by Fabre, et al.[65]. Bierlein and Arweiler's work on the electro-optic properties of KTP clearly indicates the usefulness of the material as an E-O modulator[37]. Bierlein, et al.[66], have made both planer and channel optical waveguides using an ion exchange process. Simple waveguide modulators were constructed.

7. FUTURE PROSPECTS

When one asks the question, " What are the main problems with KTP use as a nonlinear optical material ? ", the answer is usually: 1) lack of crystal size, and 2) skepticism about the damage threshold. As a result of the work by Bordui and the continued work by Philips Laboratories and Ferroxcube, the crystal size problems are becoming a non-issue. The attainment of 14 mm cubes by Loiacono with the possibility of generating SHG pieces as large as 20x20x10 mm as the development process provides improvement should satisfy the need for large aperture KTP.

The resolution of the damage problem is less straightforward, however. Users of KTP for low power applications have no damage problems to worry about. KTP becomes a problem under conditions of high power. Discoloration streaks in the bulk of the material are evident after cumulative exposure to incident high power radiation. The origin of this damage is not presently known. Some KTP crystals don't damage at all. Others damage after short exposure. Some property of the material (possibly related to it's growth process) is responsible for this type of behavior. Once the answer is in hand, the crystal growth can be adjusted to eliminate the problem. Work is presently in progress at many research institutions for the purpose of hurdling this obstacle. It is expected that many new applications for KTP will surface now that larger sizes are available and again once the damage situation is resolved.

8. REFERENCES

1. D.A. Kleinman, Phys. Rev. 128(4), 1761 (1962).
2. J.E. Bjorkholm, Phys. Rev. 142(1), 126 (1966).
3. D.A. Kleinman, A. Ashkin, and G.D. Boyd, Phys. Rev. 145(1), 338 (1966).
4. R.W. Minck, R.W. Terhune, and C.C. Wang, Appl. Opt. 5(10), 1595 (1966).
5. G.D. Boyd and D.A. Kleinman, J. Appl. Phys. 19(8), 3597 (1968).
6. S.E. Harris, Proc. of the IEEE 57(12), 2096 (1969).
7. R.G. Smith, IEEE J. Quantum Electron. 6(4), 215 (1970).
8. M.L. Ouvrard, Comptes Rendus 111, 177 (1890).
9. R. Masse and J.C. Grenier, Bull. Soc. Fr. Mineral. Cristallogr. 94, 437 (1971).
10. J.D. Bierlein and T.E. Gier, U.S. Patent No. 394323, April 6, 1976.
11. T.E. Gier, U.S. Patent No. 4231838, Nov. 4, 1980.
12. T.E. Gier, U.S. Patent No. 4305778, Dec. 15, 1981.
13. F.C. Zumsteg, J.D. Bierlein, and T.E. Gier, J. Appl. Phys. 4(11), 4980 (1976).
14. I. Tordjman, R. Masse, and J.C. Guitel, Zeit. fur Krist. 139, 103 (1974).
15. D. Cai, C. Huang, D. Shen, and Z. Yang, Gweisuanyan Xuebao 14(3), 257 (1986).
16. I.V. Voloshina, R.G. Gerr, M.Yu. Antipin, V.G. Tsirel'son, N.I. Pavlova, Yu.T. Struchkov, R.P. Ozerov, and I.S. Rez, Sov. Phys. Crystallogr. 30(4), 389 (1985).
17. A.A. Ballman, H. Brown, and D.H. Olson, J. Crystal Growth 75, 390 (1986).
18. J.C. Jacco, G.M. Loiacono, M. Jaso, G. Mizell, and B. Greenberg, J. Crystal Growth 70, 484 (1984).
19. G. Gashurov and R.F. Belt, "Growth of KTP", in Tunable Solid State Lasers for Remote Sensing, pp 119-120, Springer-Verlag, New York (1985).
20. D. Shen and C. Haung, Prog. Crystal Growth and Charact. 11, 269 (1985).
21. Y. Liu, B. Xu, J. Han, X. Liu, and M. Jiang, Zhongguo Jiguang 13(7), 438 (1985).
22. G.S. Damazyan, S.Kh. Esayan, and A.L. Manukyan, Sov. Phys. Crystallogr. 31(2), 239 (1986).
23. R.A. Laudise, R.J. Cava, and A.J. Caporaso, J. Crystal Growth 74, 275 (1986).
24. P.F. Bordui, J.C. Jacco, G.M. Loiacono, R.A. Stolzenberger, and J.J. Zola, J. Crystal Growth 84, 403 (1987).
25. G.M. Loiacono, Philips Laboratories/Briarcliff, private communication (1988).
26. N.I. Pavlova, V.M. Garmash, G.B. Sil'nitskaya, N.P. Stekol'shchikiova, and V.A. Gerken, Sov. Phys. Crystallogr. 31(1), 87 (1986).
27. V.M. Garmash, V.V. Lemeshko, V.V. Obukhovskiy, N.I. Pavlova, and I.S. Rez, Ukrian. Fiz. Zh. 31(9), 1410 (1986).
28. J. McKinlay, Philips Laboratories/Briarcliff, private communication (1985).
29. Y.S. Liu, D. Dentz, and R. Belt, Opt. Letters 9(3), 76 (1984).
30. T.E. Gier and F.C. Zumsteg, "KTP Crystals for Second Harmonic Generation", Final Report, AFAL-TR-78-208, Contract No. F33615-77-C-1131, E.I. du Pont de Nemours and Co., Wilmington, DE, Dec. 1978.
31. V.A. Kalesinskas, N.I. Pavlova, I.S. Rez, and J.P. Grigas, Litov. Fiz. Sb. 22(5), 87 (1982).
32. V.K. Yanovskii and V.I. Voronkova, Sov. Phys. Solid State 27(7), 1308 (1985).
33. V.K. Yanovskii and V.I. Voronkova, Phys. Stat. Sol. (a) 93(2), 665 (1986).
34. C. Huang, Wuli 15(5), 281 (1986).
35. G.M. Loiacono, Philips Laboratories/Briarcliff, private communication (1985).
36. V.K. Yanovskii, V.I. Voronkova, A.P. Leonov, and S.Yu. Stefanovich, Sov. Phys. Solid State 27(8), 1508 (1985).
37. J.D. Bierlein and C.B. Arweiler, Appl. Phys. Lett. 49(15), 917 (1986).
38. M.K. Rodionov, N.P. Evtushenko, and I.S. Rez, Ukrian. Khim. Zh. 49(1), 5 (1983).
39. J.C. Jacco, Mat. Res. Bull. 21(10), 1189 (1986).
40. M.K. Rodionov, V.I. Petrenko, N.P. Evtushenko, V.V. Alekseev, and I.S. Rez, Zh. Prikl. Spektrosk. 46(1), 95 (1987).
41. B. Wyncke, F. Brehat, J. Mangin, G. Marnier, M.F. Ravet, and M. Metzger, Phase Transitions 9, 179 (1987).
42. G.A. Masse, T.M. Loehr, L.J. Willis, and J.C. Johnson, Appl. Opt. 19(24), 4136 (1980).
43. M.S. Slobodyanik, G.D. Byalkovskii, and V.V. Skopenko, Dop. Akad. Nauk Ukr. RSR, Ser. B: Geol., Khim. Biol. Nauki, (6), 53 (1985).

44. G.A. Kourouklis, A. Jayaraman, and A.A. Ballman, Solid State Comm. 62(6), 379 (1987).

45. W.M. Theis, G.B. Norris, M.D. Porter, Appl. Phys. Lett. 46, 1033 (1985).

46. F. Ahmed, R.F. Belt, and G. Gashurov, J. Appl. Phys. 60(2), 839 (1986).

47. T.Y. Fan, C.E. Huang, B.Q. Hu, R.C. Eckardt, Y.X. Fan, R.L. Byer, and R.S. Feigelson, Appl. Opt. 26(12), 2390 (1987).

48. R.F. Belt, G. Gashurov, and Y.S. Liu, "KTP as a Harmonic Generator for Nd:YAG lasers", in Laser Focus 21(10), 110 (1985).

49. R.A. Stolzenberger, to be published in Appl. Opt. (1988).

50. T.A. Driscoll, H.F. Hoffman, R.E. Stone, and P.E. Perkins, J. Opt. Soc. of America B 3, 683 (1986).

51. A. L. Aleksandrovskii, S.A. Akhmanov, V.A. D'yakov, N.I. Zheludev, and V.I. Pryalkin, Sov. J. Quantum. Electron. 15(7), 885 (1985).

52. J.Q. Yao and T.S. Fahlen, J. Appl. Phys. 55(1), 65 (1984).

53. R.A. Stolzenberger, Philips Laboratories/Briarcliff, private communication (1988).

54. V.V. Lemeshko, V.V. Obukhovskiy, A.V. Stoyanov, N.I. Pavlova, A.I. Pisanskiy, and P.A. Korotkov, Ukrain. Fiz. Zh. 31(11), 1746 (1986).

55. A.P. Leonov, V.I. Voronkova, S.Yu. Stefanovich, and V.K. Yanovskii, Sov. Tech. Phys. Lett. 11(1), 34 (1985).

56. P.F. Bordui, Crystal Technology, Inc., private communication (1988).

57. V.I. Voronkova, R.S. Gvozdover, and V.K. Yanovskii, Pis'ma Zh. Tekh. Fiz. 13(15), 934 (1987).

58. J.D. Bierlein and F. Ahmed, Appl. Phys. Lett. 51(17), 1322 (1987).

59. G.M. Loiacono and R.A. Stolzenberger, submitted to Appl. Phys. Lett. (1988).

60. N. Menyuk, MIT Lincoln Laboratory Solid State Research, Quarterly Technical Report: 1 May - 31 July, 10 (1985).

61. T.S. Fahlen and P. Perkins, "Material and Medical Applications Using a 20W Frequency Doubled Nd:YAG Laser", in Digest of CLEO, 138 (1984).

62. D.N. Dovchenko, V.A. D'yakov, V.I. Kuznetsov, V.I. Pryaklin, and A.V. Simonov, Izv. Akad. Nauk SSSR., Ser. Fiz. 51(2), 259 (1987).

63. V.M. Garmash, G.A. Ermakov, N.I. Pavlova, and A.V. Tarasov, Sov. Tech. Phys. Lett. 12910), 505 (1986).

64. J.C. Baumert, F.M. Schellenberg, W. Lenth, W.P. Risk, and G.C. Bjorklund, Appl. Phys Lett. 51(26), 2192 (1987).

65. C. Fabre, E. Giacobino, S. Reynaud, and T. Debuisschert, Proc. SPIE-Int. Soc. Opt. Eng., 701(Eur. Conf.Opt., Opt. Syst. Appl., 1986), 488.

66. J.D. Bierlein, A. Ferretti, L.H. Brixner, and W.Y. Hsu, Appl. Phys, Lett. 50(16), 12 (1987).

Low temperature hydrothermal growth of KTiOPO$_4$(KTP)

R.F. Belt and G. Gashurov

Airtron Division, Litton Systems, Inc.
200 East Hanover Avenue
Morris Plains, New Jersey 07950

R.A. Laudise

AT & T Bell Laboratories
600 Mountain Avenue
Murray Hill, New Jersey 07974

ABSTRACT

KTP possesses superior properties in its use as a nonlinear optic material with Type II phase matching. It is not hygroscopic, has a large effective nonlinear coefficient, excellent optical damage resistance, small beam walk-off angle, and large thermal and angular bandwidths. These combined factors have dictated its choice as a second harmonic generator when compared to previous materials. Ba$_2$NaNb$_5$O$_{15}$ was difficult to grow, KH$_2$PO$_4$ is hygroscopic, LiO$_3$ has small spectral bandwidth, and LiNbO$_3$ suffers optical damage easily. Even the recent β-BaB$_2$O$_4$ has a larger walk-off angle, smaller angular acceptance, and a smaller d coefficient. In addition, the larger acceptance angle for KTP results in higher output for miniature configurations at low power input. KTP has now found new applications for optical waveguiding, sum frequency mixing, and parametric oscillators. The applications requiring large size have been limited because of the difficulty of growing single crystals.

The use of high temperature and pressure aqueous solvents (hydrothermal crystallization) or the use of high temperature molten salt solvents limits the size of useful crystals. Hydrothermal crystal growth at a lower temperature offers potential advantages for size, cost and quality. Recent laboratory experiments suggested that growth may be performed near 400° C. In this paper, we report the scale up of this method to results in commercial sized activities. Our preliminary data indicate that favorable growth can be maintained at temperatures of 100° C lower than previously. We have examined selected physical properties of our low temperature crystals and compared them to those from normal growth. All of our measurements indicate a KTP crystal fully comparable to previous samples. Our process represents a major advance in scale up of growth technology for this important material.

1. INTRODUCTION

The compound potassium titanyl phosphate (KTiOPO$_4$ or KTP) has become one of the most practical materials available for nonlinear optics. Single crystals were grown first by a hydrothermal method[1] employing phosphate solutions at 850° C and 30000 psi. Since this method is not suitable for crystals of several cm size, a comprehensive effort was devoted to systems which employed new mineralizers, less severe growth conditions, and commercial autoclaves. This work[2,3] culminated in a reliable process for producing high quality single crystals which were utilized for frequency doubling of Nd:YAG. All of the commercially grown material has been grown by this method and these crystals have become the standard of the industry. Such material is grown at 600° C and 25000 psi. With 1.5 x 18 inch cylindrical autoclaves, oriented crystals for Type II phase matching can be obtained in sizes up to 7 x 7 x 10 mm^3.

For some contemplated military uses which involve larger apertures of 1 cm^2 and high power fundamental laser sources, it would be expedient to prepare at least a 1 cm^3 of a properly oriented KTP cube. This goal has inspired some workers to pursue alternate growth methods. Thus, flux growth from phosphates[4] and tungstates[5] has been reported with varied success. For the last several years, Chinese workers have grown material by flux procedures[6,7] from phosphates. Sizes of 1-2 cm have been shown[8] but quality was undetermined. Some hydrothermal work[9] has also been attempted in China with KF mineralizers. The flux procedures developed with a heat pipe based furnace system[10] have yielded 10 x 10 x 7 mm^3 plates. A good quality verification has not been made. In this paper we will not dwell on the excellent physical properties of KTP as a nonlinear material. A good review of many of these has been assembled[11] for hydrothermally grown material. Similar data were collected for flux grown material of Chinese origin[12].

In recent years several new applications for KTP have become prominent in addition to frequency doubling. For example, optical wave guiding[13] and parametric generation[14] have been examined carefully and KTP crystals offer several advantages. It is becoming more attractive to ensure a steady supply of high quality, low cost, and large sizes of KTP. In 1986 the hydrothermal phase relations and solubility[15] were examined extensively in high concentrations of KH_2PO_4. It was demonstrated on a laboratory scale in 75 cm^3 platinum lined Morey type closure vessels, that KTP could be grown at lower temperatures and pressures where inexpensive steel alloys are serviceable. Since this represented a major advance in KTP growth, we deemed it necessary to perform similar experiments in our commercial vessels of 10X increased volume. This report describes our preliminary results on hydrothermal growth at the lower temperatures and compares physical properties of the KTP to the normally grown materials.

2. EXPERIMENTAL TECHNIQUES

2.1 Material preparation

The chemical components used in the KTP hydrothermal synthesis are KTP nutrient and an aqueous solution of K_2HPO_4 and a small amount of KNO_3. The KTP nutrient is prepared according to the following reaction:

$$KH_2PO_4 + TiO_2 \rightarrow KTiOPO_4 + H_2O\nearrow . \tag{1}$$

An equimolar mixture of 99.99 % pure TiO_2 and 99.9 % pure KH_2PO_4 is mixed thoroughly in a blender and prefired in a platinum crucible at 500° C. The temperature is raised to about 1200° C and the melt is kept at temperature for 12-16 hours. The molten material is then rapidly quenched by pouring it onto an aluminum slab which acts as a heat sink.

The mineralizer is 2 molar K_2HPO_4 solution. To prevent the reduction of constituents the solution is made 0.125 molar in KNO_3. Both K_2HPO_4 and KNO_3 are 99.9 % pure and the resistivity of water used is around 15 MΩ cm.

2.2 Autoclaves and liners

The autoclaves used were made of René 41; the major components of this alloy are nickel (50 %), chromium (20 %), cobalt (10 %), and molybdenum (10 %), with iron not exceeding 5 %. The autoclave dimensions were 3 inches O.D., 1.5 inch I.D., and 21 inches in length. To prevent contamination, the growth charge is placed inside an hermetically sealed gold liner (99.99 % pure) provided with a ladder for KTP crystal seeds and a baffle. The liner is 14.25 inch long, 1.45 inch O.D. and 1.375 inch I.D. The effect of the baffle, which is a perforated gold disc, is to separate the vessel into two nearly isothermal regions, one for dissolving and the other for growing. The baffle open area was about 5 %. A drawing of the system is given in Figure 1.

2.3 Heating arrangement

The autoclave is equipped with seven band heaters, each of 500 watts (See Figure 2). For the purpose of temperature control the heaters are arranged in three zones: the bottom zone contains three heaters, and two additional zones (middle and top) each consist of two heaters. The autoclave was satisfactorily insulated by firebricks (on bottom), Koawool and vermiculite. The insulation is surrounded by a double-wall steel pipe. The pipe itself is covered by a double-wall steel cover.

2.4 Pressure and temperature control

As in most hydrothermal processes, the pressure inside the autoclave is a function of temperature and degree of fill. The average temperature is nearly the same in all our hydrothermal runs. The degree of fill inside the liner varied, but most of the time was adjusted to 80 %. Enough deionized water is added to the autoclave to achieve approximately equal pressures inside and outside the liner. Of course, since gold is soft and malleable, the liner can tolerate small unbalanced isobaric conditions which are established. The initial external and internal degrees of fill must be determined empirically for correct balance of pressure.

The autoclave is connected to a pressure gauge via a special high pressure 3-way (2-way on pressure) valve. The pressure gauge is provided with two adjustable electrical contacts which activate an alarm system in case the pressure exceeds or falls below preset limits. The autoclave pressure can be regulated by either adding water to the autoclave (using an air driven pump) or withdrawing water (releasing the 3-way valve). 21,000-22,000 psi may be considered a typical operating pressure for KTP runs.

In hydrothermal growth experiments temperature is measured necessarily on the outside wall of the autoclave. It is reasonable to assume that the autoclave wall, the water outside the liner, and the wall of the liner will tend to smooth outside temperature profiles. The liner contents will also alter the temperature gradient established on the outside wall. Thus, one should keep in mind that measured temperature profiles do not represent actual temperature distributions within the liner.

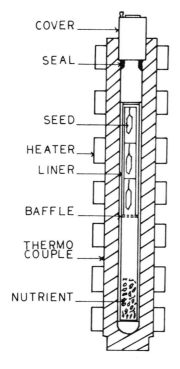

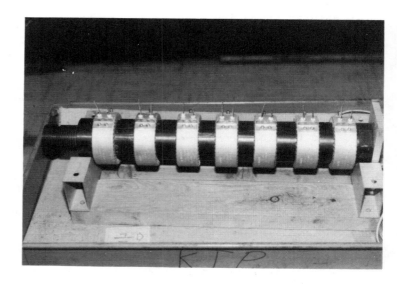

Figure 1. Drawing of autoclave and contents

Figure 2. Photograph of autoclave with band heaters in place; electric leads are not attached.

As it was mentioned in section 2.3, the seven autoclave heaters are arranged into 3 zones. Each zone is controlled by a three-mode analog controller, the three controllers being combined as a "master-slave" configuration. This means that the "slave" controllers (regulating the middle and top zones) maintain preset temperature differences relative to the temperature set on the master controller. The total stability of the controller is 0.3° C; this includes line voltage, ambient zero, ambient span, and cold junction stabilities.

The control system incorporates a temperature trimming facility for each individual heater which provides an improved control over the temperature profile within a given zone. The temperature trimming is accomplished by regulating the power output of each heater.

2.5 Run Procedure

The liner is first filled with KTP nutrient and the baffle-ladder assembly with seeds is lowered into the liner. The liner is then welded shut except for an open fill spout at top. The flux solution is added through the spout and the spout is then welded shut. The liner's volume is measured, the liner is checked for leaks and then inserted into the autoclave (seal surfaces lapped and cleaned). Pure water is added to the volume between liner and autoclave. The autoclave is then sealed and lowered into the shielded furnace. The system is brought to operating temperature and desired temperature gradients are established by adjusting controller setpoints. Pressure is adjusted by pressurizing or bleeding the system. At the end of a run the system is returned to ambient T-P by shutting off power. The autoclave is opened, and the liner removed. The liner is then cut open and the crystals are removed, washed and measured.

2.6 Measurements

Crystals were oriented and cut for frequency doubling of the fundamental wave which, in most cases, was 1.06 µm using type II phase matching. The sliced KTP cubes were polished to meet the standard 10-5 surface quality specification. The flatness was maintained at $\lambda/4$ or better, wedge and parallelism was less than 20 seconds and wavefront distortion was less than 1/2 fringe per 5 mm. The end faces of KTP cubes were antireflection coated to reduce reflection losses. The particular coatings are application dependent. However, in all cases the reflectance of the coated cube was 0.25 % or less at the operating wavelength.

Optical transmission measurements of low-temperature KTP were made by Dr. J.D. Bierlein of du Pont and its absorption in the range of 0.35 μm to 4.5 μm was found to be indistinguishable from the absorption of the high-temperature KTP. The electro-optic coefficients were also found to be the same for both types of KTP. The electrical conductivity of low-temperature KTP, however, was lower than that of conventional KTP. Preliminary damage studies carried out by Dr. Vanherzeele indicate that the damage threshold of the low-temperature KTP may be comparable to that of its high-temperature counterpart.

Conversion efficiency measurements of the second harmonic generation at 1.06 m (Type II) were made with a Q-switched Nd:YAG laser (Quantronix model 114R). The laser was operated in the TEM_{00} mode with a pulse width of 75 ns and a repetition rate of 40 Hz. The conversion efficiency of 5 mm long low-temperature KTP cubes at about 50 MW/cm^2 incident intensity was the same as that of high-temperature KTP cubes - about 35 %. The phase-matching angle and angular bandwidth were also the same for both types of KTP, $\sim$24° to x-axis in xy-plane and $\sim$8 mrad.cm, respectively.

3. RESULTS

3.1 Crystal growth and growth rate

A series of growth runs was made in which the nutrient chamber (bottom) was maintained at 475° C and the topmost position of the growth zone at about 430-435° C. Because of the relatively small size of the autoclave, the growth zone could not be maintained at isothermal conditions. The region in the growth zone just above the baffle was about 10° C below and the top of the growth zone was 40-45° C below the nutrient zone. The results of a typical low-temperature growth run (#212) will be described briefly. In this run the growth chamber contained four seed crystals. The seed orientation was (011), i.e. the seed was a plate, the large surface of which was parallel to the crystallographic (011) face. The seed dimensions were 1-1.5 mm (along the direction perpendicular to the (011) face) x 20 mm (along the <100> direction) x 30 mm (along the direction perpendicular to <100> and parallel to (011). The temperatures at the seed positions relative to the nutrient temperature (measured on the outside of the autoclave) and the corresponding growth rates are listed in Table I.

Table I. Seed Gradients and Growth Rates

Seed	ΔT, °C	Growth Rate, mm/side/week
1 (top)	47	1.13
2	44	1.05
3	31	0.91
4 (bottom)	11	0.61

The growth rates of the three top crystals compare quite favorably to those of the conventional hydrothermal KTP growth procedure in which the nutrient temperature is 590° C and the growth temperature about 575° C. The crystal quality of the low-temperature KTP crystals was generally comparable to good quality conventionally grown KTP.

The pressure-temperature relations and the effects of temperature, temperature gradient and degree of fill on the KTP growth rate have been studied by Laudise and his coworkers and the reader is referred to their paper. In the present study the growth temperature was generally 475° C and the temperature gradient about 40° C. The internal degree of fill of 80 % was used in most runs and appears to be optimum for the temperature conditions used in the present study.

3.4 Quality of crystals

In most crystal growth processes employing seeds, the quality of newly grown material is a direct function of existing defects which propagate from the seed. In order to assess the functional results of the low temperature process variables, we employed the same quality seeds whose origin was our high temperature process. These seeds were derived from one of our best growth runs. They are selected on the basis of optical clarity, microscopic examination for any inclusions, absence of cracks or veils, and a He-Ne laser examination for light scattering. The normal orientation of our seeds is (011). Plates are cut in an area of 3-4 cm^2 and a thickness of 1-2 mm. The seed plates are oriented to ±2°, polished with several μm sized abrasives, and finally etched before use. A small hole is drilled in the seed to insert a gold wire for holding it in the ladder.

In Figure 3 we show a crystal of good quality grown at the low temperature hydrothermal conditions. It should be mentioned first that we have observed no fundamental change of morphology. This might be expected on the basis of the chemical composition of the mineralizers which are essentially K^+ and PO_4^{3-} for each system. It also indicates that the transporting ion species are equal or closely associated. In many water soluble solution growing procedures, foreign ions may change growth rates of particular planes or directions, morphology, and impurity absorption. We have observed none of these and our standard passive quality tests indicate similar material from the high and low temperature material. Our chemical analyses, X-ray data, and other physical tests confirm the identity of the products. Low temperature grown crystals have been carried through all of our normal processing steps of cutting, polishing, fabrication, and AR coating for the Type II phase matching configurations. No differences are noted. One of the more demanding passive optical tests is that of the Twyman-Green interferometry. We have obtained excellent results with many crystals which prove that the internal quality is excellent.

3.5 Physical tests

One of our main concerns has been the reproducibility and identity of the nonlinear properties of high and low temperature material. To this end we have accumulated several positive results in active operation. KTP has particular applications for frequency doubling using the Type II phase matching configuration. We have taken materials from regular runs, processed them similarly, and measured conversion efficiencies under the same conditions with a Quantronix Model 114 Nd:YAG laser. Figure 4 shows a plot of green output power at 532 nm as a function of angle ϕ. The conversion efficiencies of these two uncoated crystals were both 44 ± 1 %. The width at half maximum was measured as 15.2 mrad for the high and 15.9 mrad for low temperature crystals. The length of each crystal was fabricated to 5.0 mm. Other crystals have given similar results to a reproducibility of ±5 % in the measurements. Thus, we conclude that this property is equivalent and minor variations in performance may be caused by internal quality. For example, six cubes were fabricated from the single crystal of a low temperature run. All cubes were tested for conversion efficiency and the results were 37 ±3 % under identical test conditions.

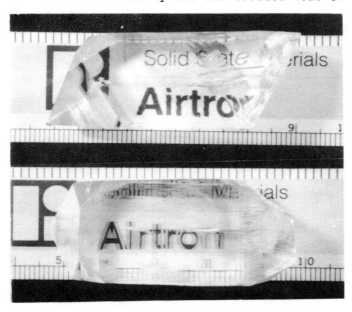

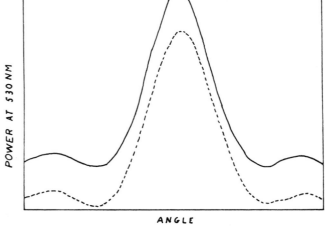

Figure 3. Composite photograph of KTP crystal from typical hydrothermal low temperature growth run. Top view is along direction parallel to (011) seed plane and bottom view is perpendicular. Some natural optical birefringerence obscures the latter.

Figure 4. Output power at 530 nm as a function of phase matching angle for Type II configuration in x-ray plane for KTP. The full line (offset vertcally for clarity) is for material at high temperature and dashed line is for low temperature.

Another critical test of crystal quality and reproducibility is the construction of waveguides[13] in KTP. For this application, substrates of (001) orientation are fabricated, polished, and Rb ions are diffused into the KTP by means of molten nitrates. A few hours at 305° C is sufficient for diffusion lengths of 4-6 μm along z+ or z- directions. In these experiments on crystal plates from both high and low temperature growth runs, similar results are obtained from the diffusions and examination of waveguiding by the m-line technique.[16] A more quantitative test is the fabrication of Mach-Zehnder type modulators from KTP. These have shown a higher efficiency and larger bandwidth than LiNbO₃ devices. Since the KTP device operation is very sensitive to material quality, these tests confirm the physical and chemical uniformity. Of particular interest is the dielectric constant since it provides the higher impedance for matching to RF drive circuitry.

One other comparison has been investigated for high and low temperature grown KTP. The test of optical damage is very critical to all of the active device uses for KTP. However, optical damage is a difficult test to reproduce, is very controversial, and should be performed with a well characterized laser beam. Preliminary results have been obtained on samples which were all coated in a similar fashion. Specifically an AR coating for 1.05 μm was depostited on the entrance face and an AR coating for 0.53 μm was deposited on the exit face for these extra cavity measurements. A Nd:YLF laser was operated in a CW mode-locked configuration with a 30 ps pulse Gaussian beam for test data. Damage criteria were based on the visible appearance of "grey tracks" within the bulk of each sample. This type of damage is typical of KTP when used with ps to ns pulse widths. The precise chemical nature of the damage has not been verified nor was the mechanism, but the observed phenomena are reproducible for KTP of any origin. Three samples of different growth conditions were examined, a high temperature hydrothermal (HTH), a low temperature hydrothermal (LTH), and a chemical flux (CF). In terms of resistance to damage, the results showed that HTH>LTH>CF. However, this was more of a qualitative assessment and further work will be reported later on more samples. As a baseline, our HTH material has been measured consistently at ∿5GW/cm for the above damage phenomena and for single shot surface damage[17] on uncoated samples. Further damage studies of our LTH material are in progress.

3.6 Other comparisons

We report here a few other interesting observations on growth of KTP via the low temperature process. The general chemical purity appears to be similar from spectroscopic tests. However, the incorporation of Fe^{3+} is nearly always lower as judged by residual color in the crystal. One might expect that temperature-solubility relations govern the uptake of Fe^{3+} regardless of the source. As all of our hydrothermal KTP contains some OH^- as determined by infrared absorption, we have found that the low temperature material is no different and may have up to an estimated 50 ppm. This absorption at 2.80 m does not affect the nonlinear properties unless one is concerned with loss at that specific wavelength. At high temperatures, KTP resembles a superionic conductor and the mobility of K^+ throughout the structure (possibly assisted by O^{2-} vacancies) increases. However, even at room temperature, the electrical conductivity shows measurable variations. We have found that low temperature grown KTP has a lower conductivity than our high temperature. This may be associated with less impurities, closer stoichiometry, or other factors related again to temperature-solubility. The formation of ferroelectric domains appears to be a function of the seed crystal and its domain structure rather than the temperature of growth. For both processes, we are well below the 936° C Curie point.

4. DISCUSSION

4.1 Phase diagram and conditions

By way of summary, we present in Table II a list of the main experimental conditions for high and low temperture growth. In these data, we have used similar nutrients, seeds, and autoclaves as described previously. It can be seen that roughly equivalent results can be attained at nearly 125° C and 4000 psi lower than in our conventional high temperature process. If we assume there is no radical change in the transporting species under the different conditions, the principal system variable is solubility of the KTP in the flux used. The high temperature flux is prepared from a mixture of the KH_2PO_4 and K_2HPO_4 to form higher phosphates. Overall it has a lower pH than the simple K HPO used in the low temperature process. It appears that TiO_2, TiO^{2+}, and certainly KTP are more soluble in solutions that are more alkaline. In fact, TiO_2 single crystals, which are very refractory, are attacked favorably by molten alkalis as KOH.

4.2 Autoclave materials

The temperature and pressure required to grow KTP by a hydrothermal process places certain restrictions on the alloys which are available for vessel construction. Obviously, it is advantageous to work with the lowest cost alloy which provides the desirable service conditions. Other factors may enter the picture; for example, in quartz growth it is important to have reaction of the vessel with the alkaline contents to form a protective layer. For KTP, where we have to incorporate the inner gold vessel to contain the highly corrosive phosphates, essentially pure water is the only compound which contacts the pressure vessel. In Table III we have collected some pertinent data for three materials which are used commonly for autoclave construction. The 4340 alloy is employed for quartz growth on a commercial scale. The A286 and Rene 41 have been used at Airtron for 20 or more years in growing a variety of crystals in lined autoclaves. In general, to work at elevated temperatures and pressures, the alloy cost, machining cost, and difficulties increase substantially. While the table does not indicate size effects, it becomes a major task to fabricate billet alloys for larger volume autoclaves, e.g. 3 inch or larger inside diameter Rene 41 autoclaves. The last three columns list the more important physical data connected with high temperature autoclave design. The ultimate tensile strength should remain high at temperature of use. Similar criteria apply to the stress rupture. The creep rate is most decisive since it determines practical life, safety features, and leak proof sealing under repetitive conditions. In normal practice, the dimensions of our autoclaves are measured as a function of hours of usage at the operating conditions. This information is translated to an expected operating life of the autoclave. Short life implies a high cost unless the autoclave can be used for another purpose under less demanding conditions.

Table II. Hydrothermal Growth Parameters of KTP

	High Temperature Method	Low Temperature Method
Temperature, °C	590	475
Temp. Gradient, °C	25	50
Pressure, psi	25,000	21,000
Flux	$K_6P_4O_{13}(12M\ PO_4^{3-})$	$2M\ K_2HPO_4$
KNO_3	0.025M	0.125M
Seed Orientation	(011)	(011)
Growth Rate in <011> Direction	1-1.5 mm/week	1-1.5 mm/week
Growth Duration	5-6 weeks	5-6 weeks
Baffle Open Area	25 %	25 and 5 %
pH of Solution (25° C)	8	11

4.3 Large system scale-up

At the present state of growing KTP by the hydrothermal method, the ultimate size of the crystal is limited solely by the autoclave size, principally its diameter. In our autoclave systems, the typical crystal size (Figure 3) is about 5.2 cm long x 1.7 cm x 1.8 cm. Fortunately, the crystal morphology "fits" the cylindrical cavity and gold liner. The latter is about 3.4 cm diameter which can only contain about 2.3 x 2.3 cm^2 dimension before the crystal intersects the wall. The seed crystal nearly bisects the finished crystal and defects are often located at the seed-crystal interface. It has been our practice to cut the commercial cubes from only the new grown material on either side of the seed. However, some (001) substrates have been fabricated favorably with the seed contained within its area. A normal scale-up is to increase the active volume of the autoclave by a factor of 10. For example, we have used autoclaves with a diameter of 3.0 inches and length of 48 inches for many crystals such as ZnO, $CaCO_3$, and Al_2O_3. One impediment with hot wall systems of the type we use, is that the Rene 41 alloys are difficult to fabricate in 6 inch diameter billets. Thus, any growth process which can tolerate less severe conditions where the A286 or 4340 alloys are serviceable is a vast improvement in technology. We will not detail our experience with these larger systems yet, but they will be the subject of another paper. It should be pointed out that even larger systems than 3 inch diameter are operable. These can vastly increase the yield of KTP at 5 mm cubes, can easily obtain 1 cm^3 cubes, and can yield (001) substrates up to 10-20 cm^2 in area. With a larger system it would be advantageous to increase the growth rate as long as quality is not compromised. This is important because of the longer growth run times for increased size of crystals. Presently, we see no major problems in running autoclaves continuously for 2-3 months.

Table III. Selected Properties of Autoclave Materials

Alloy Code	Description	Composition (%)	Cost ($/lb)	Machinable	Ultimate Tensile Strength at 538°C (1000 psi)	Stress Rupture at 650°C, 1000 hr (1000 psi)	Creep Rate at 650°C, .5% in 1000 hr (1000 psi)
4340	Ultra high strength	.4 C, .8 Mn, .2 Si, .8 Cr, 1.8 Ni, .2 Mo Fe balance	1	easy	240[a]	50[a]	10[b]
A286	Iron base Superalloy	14 Cu, 25 Ni, 1.2 Mo, 2.2 Ti, .3 V, .2 Mn, .2 Si, Fe Bal.	8	moderate	131	46	30
René/ U41	Nickel base alloy	18.7 Cr, 9.8 Mo, 10.8 Co, .15 Al, 3.1 Ti, 1.4 Fe, Ni Bal.	60	difficult	187	96	73

a) At 450° C
b) At 540° C

5. CONCLUSIONS

The hydrothermal method has become an important technique for the growth of the nonlinear optical material KTP. Present commercial procedures require the use of expensive non-ferrous alloys to attain the growth conditions of 600° C and 25000 psi pressure. We have investigated a lower temperature and pressure laboratory process which employs a mineralizer solution higher in pH and effective PO_4^{3-} concentrations. The application of this process has been scaled up to our commercial size autoclaves. We have obtained favorable growth results at 475° C and 21000 psi to yield equivalent growth rates, size, and crystal quality. The quality has been confirmed by normal passive processing tests and several active nonlinear experiments. Our data confirm the validity of the lower temperature process and identity of physical properties of the crystals. The less severe growth conditions have a wide impact on autoclave alloys, attainable crystal size, quality, and cost. Further investigations are planned in larger systems.

6. ACKNOWLEDGMENTS

The authors wish to express gratitude to their respective parent companies for permission to publish these results. We are indebted to Dr. John Bierlein of E.I. du Pont Laboratories for communication of data on waveguide devices and optical damage.

7. REFERENCES

1. F.C. Zumsteg, J.D. Bierlein and T.E. Gier, J. Appl. Phys. 47, 4980 (1976).
2. R.F. Belt and L.E. Drafall, Paper given at ACCG-5, July, 1981, San Diego, CA.
3. R.F. Belt, L.E. Drafall, G. Gashurov and Y.S. Liu, "Nonlinear Optical Materials for Second Harmonic Generation", AFWAL-TR-84-1074, May, 1984.
4. J.C. Jacco, G.M. Laiacono, M. Jaso, G. Mizell and B. Greenberg, J. Crystal Growth 70, 484 (1984).
5. A.A. Ballman, H. Brown, D.H. Olson and C.E. Rice, J. Crystal Growth 75, 390 (1986).
6. C.E. Huang and D.Z. Shen, J. Synthetic Crystals 14, 141 (1985). In Chinese.
7. D.F. Cai, C.E. Huang, D.Z. Shen, and Z.T. Yang, J. Chin, Silicate Soc., 14, 257 (1986). In Chinese.
8. D. Cai and Z. Yang, J. Crystal Growth 79, 974 (1986).
9. S. Jia, P. Jiang, H. Niu, D. Li, and X. Fan, J. Crystal Growth 79, 970 (1986).
10. P.F. Bordui, J.C. Jacco, G.M. Laiacono, R.A. Stolzenberger and J.J. Zola, J. Crystal Growth 84, 403 (1987).
11. R.F. Belt, G. Gashurov and Y.S. Liu, Laser Focus 21, 110 (1985).
12. T.Y. Fan, C.E. Huang, B.Q. Hu, R.C. Eckhardt, Y.X. Fan, R.L. Byer and R.S. Feiglson, Appl. Optics 26, 2390 (1987).
13. J.D. Bierlein, A. Ferretti, L.H. Brixner and W.Y. Hsu, Appl. Phys. Letters 50, 1216 (1987).
14. H. Vanherzeele, J.D. Bierlein and F.C. Zumsteg, Appl. Optics. To be published.
15. R.A. Laudise, R.J. Cava and A.J. Caporaso, J. Crystal Growth 74, 275 (1986).
16. P.K. Tien, R. Ulrich and R.J. Martin, Appl. Phys. Letters 14, 291 (1969).
17. F. Ahmed, Appl. Optics. To be published.

Isomorphous substitution in potassium titanyl phosphate

Richard H. Jarman and Stephen G. Grubb

Amoco Technology Company
P.O. Box 400
Naperville, Illinois 60566

ABSTRACT

Potassium titanyl phosphate (KTP) is a very important material for use in nonlinear optics. In particular, its nonlinear coefficients and refractive indices are well suited for efficient generation of the second harmonic of the $1.06\mu m$ line of the Nd:YAG laser. There is growing interest in manufacturing compact diode-pumped solid-state lasers with wavelengths in the visible region. The nonlinear crystal is a key component in these devices. Although it is possible to obtain blue radiation by sum frequency generation of the diode wavelength and $1.06\mu m$, the non-critical point for type I phase matching for the second harmonic falls at around 990 nm. KTP is, therefore, unsuitable for obtaining wavelengths below 495 nm by second harmonic generation.

There are many analogs of KTP in which any of the potassium, titanium or phosphorous are isomorphously replaced by other elements. For example, K can be replaced by Rb, Tl or NH_4, and P by As. The properties of several of these compounds are known. Very little is known about the compound in which Ti is replaced by Sn. We describe the growth of a series of crystals $KTi_{1-x}Sn_xOPO_4$ from high temperature solution, and the characterization of the optical properties of these materials by powder methods.

1. INTRODUCTION

Potassium titanyl phosphate, $KTiOPO_4$ (KTP), has important applications as a nonlinear component in lasers and electro-optic systems. Currently, the most common use of single crystals of KTP is in high power Nd:YAG lasers for the production of green 532 nm radiation by second harmonic generation (SHG) of the $1.06\mu m$ fundamental. More recently, this application has been extended to the production of compact diode pumped visible lasers.[1] Two important properties of the nonlinear crystal are the magnitude of the nonlinear optical coefficient (d_{ijk}) and the birefringence. The former will determine the maximum efficiency of the frequency conversion process, while the latter will determine the required polarization and direction of propagation of the input beam or beams. A direct consequence of the dispersion of the refractive indices is that an upper limit is placed on the frequency that can be doubled. For KTP and type II phase-matching, that is 990 nm.[2,3] As a consequence of this, blue light is not obtainable by SHG using KTP. However, it has been demonstrated that a wavelength of 459 nm is obtained by frequency addition of the 1.06 μm output of Nd:YAG and the 809 nm of the laser diode.[4]

Many isomorphs of KTP exist with substitution on any of the K, Ti or P sites. By direct synthesis, Rb,Tl and NH_4 can be substituted for K[5,7], Sn for Ti[8] and As for P.[5,6,9] It has also been demonstrated that other isomorphs can be prepared by post-synthesis treatment of the material. Na can be exchanged for K by immersion of KTP in molten $NaNO_3$ for an extended period.[10] In the case of NH_4TiOPO_4 (NTP), half of the ammonia can be removed to form $NH_4H(TiOPO_4)_2$ (NHTP).[11] Furthermore, this compound then absorbs water topotactically to form $NH_4H_3O(TiOPO_4)_2$ (NOTP).[11]

It is to be expected that both d_{ijk} and the refractive indices will be affected by the substitutions. In the series $MTiOAsO_4$ (M = K,Rb,Tl), the SHG signal at 532 nm obtained from powders was observed to decline significantly as the atomic number of M increased.[9] In the NTP system, dramatic changes in SHG were observed upon removal of NH_3 and addition of H_2O, which were correlated with changes in the nature of the oxygen coordination around Ti.[11] In none of these compounds, have changes in phase-matching been investigated. To date, no information has been available regarding isomorphs containing Sn in place of Ti. In this paper, we investigate the synthesis of a series of materials containing Ti and Sn and characterize their NLO properties using powder techniques.[12-14] We describe measurements as a function of particle size and wavelength to determine any changes in the phase-matching points as a function of composition.

2. EXPERIMENTAL

Samples of $KTi_{1-x}Sn_xOPO_4$ were prepared by two methods: solid state reaction and flux growth. In the former case, stoichiometric mixtures of TiO_2, SnO_2, and KH_2PO_4 were ground together thoroughly and fired at 950°C for 16 hours in alumina crucibles. For flux growth, TiO_2, SnO_2, K_2CO_3 and KH_2PO_4 were mixed together to give the following mole fractions in terms of oxides: 0.15 MO_2 (M = Ti + Sn), 0.47 K_2O, 0.38 P_2O_5. The mixtures were heated in platinum crucibles at 400°C and then raised to 1100°C for several hours prior to slow cooling from 950°C to 680°C at 2°C per hour. The crucibles were then cooled rapidly to room temperature. Excess flux was removed by dissolution in boiling water. Starting compositions are shown in Table 1. All the products were examined by powder X-ray diffraction. Lattice parameters were

Table 1

Preparative and Structural Data for $KTi_{1-x}Sn_xOPO_4$

x	Preparation	$a/\text{Å}$	$b/\text{Å}$	$c/\text{Å}$	$V/\text{Å}^3$
0.0	Flux growth*	12.808(1)	10.575(2)	6.406(1)	868
0.1	Flux growth	12.810(2)	10.573(3)	6.408(2)	868
0.1	Solid state**	12.844(1)	10.591(1)	6.420(1)	873
0.33	Flux growth	12.979(2)	10.676(6)	6.483(4)	898
0.33	Solid state	12.906(2)	10.618(2)	6.468(3)	886
0.5	Flux growth	13.049(3)	10.670(1)	6.488(1)	903
0.5	Solid state	12.929(4)	10.626(2)	6.447(3)	886
1.0	Solid state	13.129(3)	10.708(4)	6.518(3)	916

* Flux has composition $0.15\ (MO_2)$: $0.47\ K_2O$: $0.38\ P_2O_5$
Slow cooling: 950-680°C, 2°C per hour

** Stoichiometric mixture heated at 950°C

refined using a least-squares iterative procedure and are shown in Table 1.

A diagram of the apparatus used for measuring the non-linear optical properties of the samples is shown in Fig. 1.

referenced to a powdered quartz standard. Several different sample configurations were compared including backscattering, collection of forward and backscattered signal with the use of a parabolic mirror, and in transmission geometry with index matching

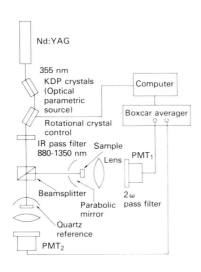

FIGURE 1

Equipment layout for powder measurements of second harmonic generation. The configuration shown is for monitoring as a function of wavelength. For measurements of SHG as a function of particle size (see text), the Nd:YAG laser is operated at 1.06 μm.

Tunable, high peak power radiation for the wavelength dependence experiments was obtained by use of an Optical Parametric Source as described by Anthon, et al.[15] Two counter-rotating type II KDP crystals are pumped by 30 psec, 355 nm pulses fron a Quantel amplified mode-locked Nd:YAG laser. The infrared portion of the spectrum from this source is broadly tunable from 880-1400 nm. The SHG signals were averaged by a Stanford Research Systems boxcar integrator

fluid. The noncritical phase matching point was found to be the same in all cases although the magnitude of the signal change when tuning through this point could differ significantly. Two experiments were performed. The first was that described by Kurtz and Perry,[12] in which the SHG output at 532 nm (using the 1.06μm line from a high power Nd:YAG laser) from a known pathlength of material is monitored as a function of particle size relative to quartz powder of the same particle size. The intensity of the SHG signal gives a semi-quantitative indication of the magnitude of the second order susceptibility, and the variation of the SHG signal with particle size is determined by the phase-matchability at $1.06 \mu m$. In the second experiment, described by Halbout et al.,[13] the SHG intensity is monitored as a function of wavelength at a fixed particle size which is greater than the coherence length. This enables the non-critical phase-matching wavelength to be determined.

3. RESULTS

3.1 Synthesis

All of the products from the solid state and flux preparations were determined from powder X-ray diffraction experiments to be highly crystalline compounds having the KTP structure. As the amount of Sn in the reaction mixture is increased, the unit cell volume increases, as shown in Table 1. The a and c parameters increase more rapidly than the b parameter. This increase is somewhat less than that expected on the basis of the ionic radii for Ti^{4+} and Sn^{4+} which are 0.68 A and 0.71 A respectively.

In the case of the solid state preparations, it was noted that a small SnO_2 impurity remained after firing, the amount of which increased with x. This implies that the actual Sn contents of the mixed phases for x = 0.33 and 0.5 are slightly lower than those predicted from the stoichiometry of the starting materials.

Comparison of the lattice parameters for the solid state products and the flux products prepared with the same nominal composition, Table 1, reveals differences in the unit cell volumes. For x = 0.1, the flux product has a smaller unit cell volume. In fact, the lattice parameters are indistinguishable from those of pure KTP, which implies that, at low concentrations, Sn is not incorporated into the lattice during crystallization. For x = 0.33 and 0.5, the unit cell volumes of the flux products are larger than those of their solid state counterparts. This may indicate preferential incorporation of Sn at higher concentrations, or may merely reflect the lower Sn contents of the solid state products.

It was observed that SnO_2 was not very soluble in the $K_6P_4O_{13}$ flux, even at 1100°C. In the preparations for x = 0.5 and 1, the solid phase did not dissolve completely even after prolonged heating at 1100°C. The conversion of the SnO_2 to the product did proceed to completion however. The particle size in these cases was much smaller than in the Ti rich samples, and crystal facets were not well developed. Lowering the SnO_2 concentration to 10 mole percent did not lead to dissolution. A flux of composition $3K_2WO_4 \cdot P_2O_5$[15] was also unable to dissolve SnO_2.

No products were obtained from this experiment.

3.2 Powder Measurements

The variation of the 532 nm of the pure KTP sample is shown as a function of particle size in Fig. 2. As expected, the intensity increases with the particle size, indicating that the material is phase-matchable at $1.06 \mu m$. The pure Sn form showed no SHG at all from $1.06 \mu m$. The intermediate forms show a systematic decrease in SHG intensity with increasing x.

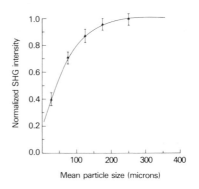

FIGURE 2

SHG intensity as function of particle size for pure KTP at 1.06 μm. Sn substituted forms show similar behavior. The shape of the curve is characteristic for phase-matching at 1.06 μm[12].

The SHG intensity as a function of wavelength is shown in Fig. 3. Points were obtained at 10 nm intervals. The intensity falls off strongly between 1010 nm and 960 nm, indicating that the non-critical phase-matching point lies somewhere near the midpoint of this region, as is well known from the refractive indices.[2,3] A similar curve is shown for the material with x = 0.5. The form of the curve is unchanged from that for pure KTP, indicating that the non-critical phase-matching point is unchanged by Sn substitution.

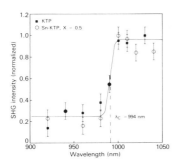

Normalized SHG intensity as a function of wavelength for KTP and $KTi_{0.5}Sn_{0.5}OPO_4$. Non-critical phase-matching point is shown at λ_C.

FIGURE 3

We conclude from these results that d_{ijk} decreases with increasing Sn incorporation in the KTP structure until it approaches zero for the pure Sn form. At the same time, the magnitude of the birefringence is not altered, although the absolute values of the refractive indices may change with composition.

It has been speculated that the large d_{ijk} for KTP arises from the presence of the chains of alternating long and short Ti-O bonds.[7] The decrease in SHG intensity when the Ti=O-Ti groups are converted into Ti-OH-Ti groups by removal of NH_3 from the NH_4 form of KTP strongly supports this assertion.[11] In the case of Sn substitution, we speculate that the Sn atom occupies a more central position in the SnO_6 octahedron than the Ti atom in the TiO_6 octahedron, although the ionic radii are very similar. The alternating short and long Ti-O bonds arise through a ferroelectric distortion of the TiO_6 octahedra, a well known phenomenon in transition metal oxides. This is accomodated by π bonding between the lone electron pairs on the oxygen atom with unoccupied 3d orbitals on the Ti atom. In the case of Sn, the bonding in the SnO_6 octahedron can be described approximately in terms of sigma bonds made up of 5s, 5p and 5d orbitals from Sn mixed with sp orbitals from the oxygen atoms. The 5d orbitals are too high in energy to be involved in π bonding with the O atoms. Therefore, the system cannot be stabilized by a ferroelectric distortion and the SnO_6 octahedron remains regular. A structural study of $KSnOPO_4$ is in progress to test this hypothesis.

4. CONCLUSION

We have prepared a range of compounds $KTi_{1-x}Sn_xOPO_4$ in which Sn is substituted for Ti in the KTP structure. With increasing x,

the magnitude of the SHG obtained from the 1.06μm pump decreased, to near zero at x = 1. The birefringence of the materials appeared to be unchanged over the range of x. We speculate that, unlike Ti there are no low lying d orbitals around the Sn atom, consequently there is no d_π-p_π bonding in the SnO_6 octahedron, and so the geometry is regular. As a result, the large non-linearity arising from the Ti=O group is lost in the Sn substituted forms.

5. ACKNOWLEDGEMENTS

We thank S.A. Gramsch and J.A. Kaduk for powder x-ray diffraction data.

6. REFERENCES

1. See for example, T. Baer, J. Opt. Soc. Am., B3, 1175 (1986).
2. T.Y. Fan, C.E. Huang, B.Q. Hu, R.C. Eckhardt, Y.X. Fan, R.L. Byer and R.S. Feigelson, Applied OPtics, 26, 2390 (1987).
3. K. Kato, IEEE J. Quantum Electronics, QE24, 3 (1988).
4. W.P. Risk, J.-C. Baumert, G.C. Bjorklund, F.M. Schellenberg and W. Lenth, J. Opt. Soc. Am., A4, 128 (1987).
5. J.D. Bierlein and T.E. Gier, U.S. Patent 3,949,323 (1976).
6. T.E. Gier, U.S. Patent 4,305,778 (1981).
7. F.C. Zumsteg, J.D. Bierlein and T.E. Gier, J. Appl. Phys., 47, 4980 (1976).
8. M.L. Ouvard, Comptes Rendus, 111, 177 (1890).
9. M. El Brahimi and J. Durand, Revue de Chim. Min., 23, 146 (1986).
10. R.H. Jarman, 11th International Symposium on the Reactivity of Solids, Princeton N.J. 1988.
11. M.M. Eddy, T.E. Gier, N.L. Keder, G.D. Stucky, D.E. Cox, J.D. Bierlein and G. Jones, Inorg. Chem., 27, 1856 (1988).
12. S.K. Kurtz and T.T. Perry, J. Appl. Phys, 39, 3798 (1968).
13. J.M. Halbout, S. Blit and C.L. Tang, IEEE J. Quantum Electronics, QE17, 513 (1981).
14. R.V. Kochikyan, V.M. Markushev, Yu.O. Yakovlev, V.R. Belan, V.F. Zolin and L.G. Koreneva, Sov. J. Quantum Electonics.
15. D.W. Anthon, H. Nathal, D.M. Guthals and J.H. Clark, Rev. Sci. Instrum., 58, 2054 (1987).
16. A.A. Ballman, H. Brown, D.H. Olson, and C.E. Rice, J. Crystal Growth, 75, 390 (1986).

ADDENDUM

The following paper, which was scheduled to be presented at this conference and published in this proceedings, was cancelled.

968-04   **Large ALON windows**
R. L. Gentilman, E. A. Maguire, Raytheon Co.

The following papers were presented at this conference, but the manuscripts supporting the oral presentations are not available.

968-09   **Review of nonlinear optical crystals for frequency conversion**
D. Eimerl, Lawrence Livermore National Lab.

968-10   **Experimental investigation of second harmonic generation in phosphorous-doped optical fibers**
J. R. Rotgé, M. E. El-Hewie, Frank J. Seiler Research Lab.

AUTHOR INDEX